Vapour Phase Synthesis of Few Alkyl Aromatics Over Ferrites Catalysts in Fixed Bed Reactor

DR. RAJEEV DIXIT

London Journals Press

Vapour Phase Synthesis of Few Alkyl Aromatics over Ferrites Catalysts in Fixed Bed Reactor

Chemistry is the Melodies you can Play on Vibrating Strings.

Michio Kaku

Dr. Rajeev Dixit

Vapour Phase Synthesis of Few Alkyl Aromatics over Ferrites Catalysts in Fixed Bed Reactor

London Journals Press (UK)
8 Jan. 2015

Copyright © London Journals Press, 2015.
All the rights reserved

1210th, Waterside Dr,
Opposite Arlington Building,
Theale, Reading
Pin: RG7-4TY
United Kingdom
E-mail: helpdesk@londonjournalspress.org

CONTENTS

Page No.

Chapter-1 Introduction   1-26

    1.1 Alkylanilines

    1.2 Alkylpyrazines

    1.3 Alkylphenols

    1.4 Alkylpyridines

    1.5 Nanocrystalline Materials

    1.6 Ferrites

    1.7 X-Ray Diffraction Technique

    1.8 Mossbauer Spectroscopy

    1.9 Surface area of catalyst by Brunauer Emmett and Teller (BET) Method

    1.10 Gas Chromatography

    1.11 Recent Literature Survey

    1.12 Reason for undertaking the present problem

    1.13 References

Chapter-2 Experimental   27-30

    2.1 The Catalytic Unit

    2.2 Preparation of ferrite catalysts

    2.3 Measurement of acidity by ammonia desorption method

    2.4 GLC analysis of reaction mixture

    2.5 Nomenclature and Sample Calculation

    2.6 The XRD recording

    2.7 SEM recordings

    2.8 Mossbauer measurements

    2.9 Atomic Force Microscopy

    2.10 ESCA measurements

Chapter-3 Methylation of aniline over Mn-Cu ferrites catalysts   31-37

    3.1 Introduction

    3.2 Experimental

    3.3 Results and Discussion

    1.4 Conclusion

    1.5 References

Chapter-4 Kinetics Studies and Mechanism Evolution of the cyclization of Ethylene diamine and propylene Glycol over Alumina supported naocrystalline Mn-ferrite 38-46

    4.1 Introduction

    4.2 Experimental

    4.3 Results and Discussion

    4.4 Reaction Mechanism

    4.5 References

Chapter-5 Kinetic Studies and Mechanism Evolution of the 118-136 Methylation of Phenol Over Al- Fe_2O_4 Catalyst

47-55

    5.1 Introduction

    5.2 Experimental

    5.3 Result and Discussion

    5.4 Reaction Mechanism

    5.5 References

Chapter - 1

I. INTRODUCTION

Basically, chemicals are derived from Plants, animals and petroleum. Petroleum and coal are the largest natural source of hydrocarbons. However, Chemical obtained from petroleum find major applications as fuels and solvents. In order to make petroleum and coal chemical more useful they need to be functionalised. Simplest derivatives of hydrocarbons are their alkyl forms. Once the hydrocarbon is alkylated, they can easily be oxidized to their acid, amide, nitrile and amine derivative. Alkyl benzenes can be oxidized to corresponding acids which find applications as medicine and polymers. Alkyl. pyridines can be oxidized to corresponding acids and amides which find application as basic drugs. However, availability of alkyl aromatics from natural source is poor and can not meet the demand of drug and polymer industries. In view of this, it becomes important to go for synthesis of these chemicals. When ever possible vapour phase production of chemicals over solid surfaces is preferred because of repeatability of the process and its eco-friendly nature. In the present thesis, production processes of some alkyl aromatics such as N-alkylated anilines, alkylated phenol and alkylated pyrazine are discussed in detail.

The thesis is presented in five chapters. Details of process, catalysts, theories, literature survey and reason for undertaking the present problem is outlined in the subsequent sections of this chapter. A detail of experimentation is presented in chapter 2. Process optimisation for N-alkylation of aniline is presented in chapter 3. Kinetics and mechanism of the vapour phase cyclization of propylene glycol and ethylene diamine to 2-methyl pyrazine is presented in chapter 4. Kinetics of alkylation of phenol is presented in chapter 5.

1.1 Alkylanilines

Alkylanilines are important intermediates for the production of dyes, herbicides and pharmaceuticals. N, N-dimethylaniline (NNDMA) is employed in the synthesis of triphenyl methane dyes such as basic green1, acid blue and basic violet. Nitration of NNDMA produces tetryl (2,4,6-trinitrophenylmethyl nitramine) used as booster explosive. 2,6-diethylaniline is starting material for synthesis of herbicide butachlor. N-alkylamines are used for synthesis of pharmaceutical compounds. When alkylation is in the benzene ring, the products are called C-alkylated anilines. Because of the delocalization of the unshared pair of electrons over the amine group into the benzene ring, the ortho and para sites of the ring become higher in electron density compared to the meta position. As a result the NH_2 group becomes ortho/para directing and aniline is expected to produce ortho and para C-alkylated product. However, when alkylation is performed by methanol over a solid catalyst, one generally obtains N-alkylated product. This can be explained due to reduced electron density on the ring because of adsorption. Catalyst surface also help in production of a carbonium ion or a positively charged species from the alcohol molecule which functions as an electrophile. The mechanism of alkylation of aniline by methanol over solid catalyst is supposed to be a consecutive reaction. Monomethyl aniline is formed first which is subsequently alkylated to N,N-dimethyl aniline. The catalytic cycle for alkylation over catalysts possessing only Bronsted sites such as ZSM-5 is shown in Fig.1.1 As usual the alcohol molecule is protonated by the Bronsted site, releases a molecule of water and is converted to a carbocation which attacks as an electrophile to the amino group of unadsorbed aniline and produces N-methyl aniline. Attack by another carbocation over N-methyl aniline produces N,N-dimethyl aniline. Protonation of aniline over Bronsted sites may reduce the yield of methyl anilines in the beginning due to formation of protonated aniline Fig. 1.2. However excess of methanol can flush away protonated aniline and yield may increase subsequently. It may be noted that the molar ratio of aniline to methanol in aniline alkylation is generally 1:6.

The catalytic cycle for aniline alkyaltion over catalysts possessing only Lewis sites such as transition metal mixed oxides is shown in Fig. 1.3. Since aniline is stronger base than alcohol, it will be preferentially adsorbed over Lewis cites, but excess of alcohol will flush it away and get adsorbed. Initial reaction between aniline and methanol will produce H_2O which will generate Brønsted sites. Further reaction between aniline and methanol will follow mechanism similar to that explained for Brønsted sites (fig. 1.4).

There are reports on alkylation of aniline in liquid phase under pressure[1], in vapour phase using oxides[2], Raney-Nickel[3], zeolites[4], AEL type molecular sieves[5], clays[6-10] and few spinels[11-14].

1.2 Alkylpyrazines

Compounds containing two nitrogen atoms in the benzene ring are categorised as diazines. Thus, three isomeric diazines having nitrogen in ortho, meta and para positions are known as pyridazine, pyrimidine and the pyrazine respectively.

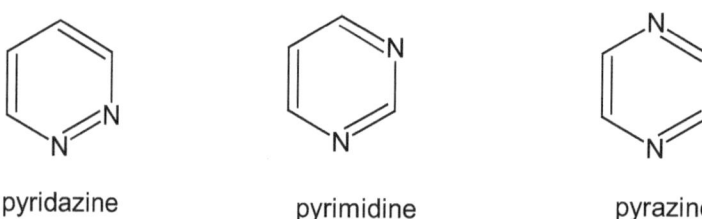

pyridazine   pyrimidine   pyrazine

Few pyridazine derivatives such as maleic hydrazide is used as a selective plant growth regulator.

Maleic hydrazide   Barbituric acid

The pyrimidine skeleton is an important component of RNA and DNA and is also present in many natural products, such as Vitamin B1 (thiamine) and many synthetic compounds, such as barbituric acid. Pyrazine derivatives find applications as flavouring materials and drugs.

pyrazinamide

2-methyl pyrazine is the starting material for synthesis of pyrazine derivatives. 2-methyl pyrazine (2-MP) is the pivotal intermediate for obtaining 2-cyanopyrazine, which on hydrolysis yields pyrazinamide, a well-known anti-tubercular drug. There are reports on the use of Palladized Zn-Cr oxide [15], ZSM-5 [16], modified copper-chromite [17], binary catalyst based on oxides of Zn and variable valence metals [18], Zn-modified ferrierite [19], Zn-modified zeolites [20] and $CuO/ZnO/SiO_2$ [21], as catalysts for synthesis of 2-MP from ethylene diamine (ED) and proplene glycol (PG).

Kulkarni et al. have reported a Rediel-Eley type mechanism of 2-MP formation from PG and ED over ZSM-5 catalyst possessing Bronsted sites. According to them, PG is adsorbed on the surface and gives rise to carbocation, which rearranges to give propylene oxide and propanaldehyde. Propylene oxide and propanaldehyde both can react with ethylene diamine to produce piperazine and 2-MP. Propylene oxide can also rearrange to give acetone. According to this mechanism, only PG is adsorbed and ethylenediamine remains in the gas phase. However, over Lewis catalysts, this mechanism does not seem to sound well instead a Langmuir-Hinselwood type of mechanism seems to be operative because ethylene diamine as well as propylene glycol seem to have equal probability of adsorption as both are used in almost equimolecular proportion. The mechanism over Lewis sites is shown in Fig.1.5

1.3 Alkylphenols

Alkylated phenols such as o-cresol and 2,6-xylenol are industrially valuable compounds, which are utilized extensively as starting materials and intermediates in agrochemical and polymer industries. The simplest method to prepare these chemicals is by methylation of phenol. Over the past three decades a variety of catalysts has been investigated for methylation of phenol[22-40]. It has been established that horizontal adsorption of phenol on acidic catalysts such as SiO_2-Al_2O_3, H-ZSM-5 and HY, promote O-alkylation, giving phenol ether[22-26], while vertical adsorption at the catalyst surface leads to ring alkylation[27]. As shown in Schemes 1 and 2 horizontal and vertical adsorptions are defined in terms of the O-C bond as his orientation for phenoxide/phenol absorbed perpendicularly to the surface. Basic catalysts such as MgO, ZnO-Fe_2O_3 promote alkylation by yielding o-cresol and 2,6-xylenol through perpendicular adsorption of phenol over the catalyst surface[28-30]. Mixed metal oxides, such as MgO-based and V_2O_5-based[32-34] compounds and Al-containing hydrotalcites[35-37], produce o-cresol and 2,6-xylenol,on varying acido-basic properties and reaction temperature. The rare earth metal oxides such as CeO_2, SnO_2 and La_2O_3 favour ring alkylation[38]. Recently, phenol methylation over ferrospinels based on Co, Ni and Zn was investigated by Sree Kumar and Sugunan[39-40].

1.4 Alkylpyridines

Methylpyridines and their derivatives are industrially important compounds, as fine chemicals, bulk drugs, insecticides, fungicides and herbicides[41-42].Increasing demands for the production of methylpyridines and their derivatives exceeds that traditionally obtained from coal tar distillation. Attempts have therefore been made to develop synthetic processes for these chemicals[43-45]. Vapour phase catalytic processes seem to be the preferred technology as this does not produce any solid waste and is environmentally friendly.

1.5 Nanocrystalline Materials

Nanopowders, controlled to nanocrystalline size (under 100nm) can show atom like behaviors which result from higher surface energy due to their large surface area and wide band gap between valence and conduction band when they are divided to near atomic size. One of the most important findings was the existence of size dependent properties. Changing the materials particle/crystal size can actually alter properties that were earlier thought to be constant for a given material. Unique surface chemistry was also attributed to nanoparticles due to the discovery that chemical reactivity is increased, as a particle is made smaller. These phenomena can effectively enhance optical, chemical, electro-magnetic properties of materials. Such materials are finding applications such as UV protection, photocatalysis, field emission displays, varistors, and functional devices, thermoelectric materials due to exceptional physical and chemical qualities.

Nanocrystalline materials, involving crystallites of 1-10 nm in dimension possess ultrahigh surface-to-volume ratio due to their fine grain sizes[46,47]. Such materials contain large number of atoms located at the edges and on the surfaces, which provide active sites for catalyzing surface reactions. Nanocrystallites further possess unique hybrid properties approaching those of atoms and molecules. Nanocrystalline processing offers a practical way of retaining the results of property manipulation on the atomic or molecular level, producing novel materials with unique size-dependent behavior, including quantum confinement effects, superparamagnetism, greater microstructural uniformity for better mechanical reliability, and high ductility and superplasticity for advanced ceramics. Over the past decade, research on nanocrystalline materials has been greatly accelerated by the advances in our ability to manipulate structures on the molecular or atomic level. However, most of the studies have been directed towards the synthesis, characterization, and application of these systems as structural and optical/electronic materials. As catalysts, nanometer-sized active clusters have been examined for a long time, but they are mainly limited to supported metal systems. Direct synthesis and successful stabilization of nanocrystalline metallic and ceramic materials have only recently been investigated in detail for some catalytic applications [48]. It would be particularly fruitful to exploit the nanocrystalline systems for their size-dependent effects in structure-sensitive reactions, whereby the catalytic activity depends not only on the number of active sites, but also on the crystal structure, interatomic spacing, and crystallite size of the catalytic material.

A feature of nanocrystalline processing can be demonstrated with the help of a study on nanocrystalline cerium oxide-based catalysts [49]. By inert gas condensation of Ce clusters followed by controlled post-oxidation, it is possible to achieve highly non-stoichiometric CeO_{2-x} [50]. The unusually high oxygen vacancy concentration possessed and retained by these nanocrystallites is associated with active super oxide surface species. Compared to the conventional high surface area CeO_2, nanocrystalline CeO_{2-x} enabled catalytic activation at a significantly lower temperature for SO_2 reduction and CO oxidation, and demonstrated superior poisoning resistance [51,52]. Nanocrystalline processing further allowed ultrahigh dispersion of components, such as that demonstrated in Cu-CeO_{2-x} nanocomposites [53]. Unique chemical and electronic synergistic effects can be obtained from such homogeneously dispersed multicomponent systems, useful towards promoting catalytic activity, especially when they are thermally stable under the reaction conditions.

1.6 Ferrites

Nanocrystalline spinels, formed in solution phase by simple adjustment of pH, posses wide scientific and technological interest among the mixed oxides. Spinel ferrites are used as magnetic materials in wide variety of applications in the electronic industry ranging from transformer materials to recording media. In catalysis, spinel structures can be found in various systems. For example, γ-Al_2O_3 is a very common support material and many other spinels such as $MnMn_2O_4$ (α-Mn_3O_4), $MnAl_2O_4$, $ZnMn_2O_4$, $CoCo_2O_4$, (Co_3O_4), $ZnCo_2O_4$, $ZnAl_2O_4$, 20% MnO / ZnO and ZnO find application as catalyst for various reactions. This seems to be a reason for large volume of literature appearing on the synthesis, characterization and applications of spinels [54-56].

The spinel oxides have the general formula AB_2O_4, where A and B are cations with oxidation states of +2 and +3, respectively. Although, the most numerous and most interesting compounds with the spinel structure are oxides, some halides and sulphide also crystallize in this structure. The spinel structure, named for the mineral spinel '$MgAl_2O_4$', is face centred cubic with a large unit cell containing eight formula units [57-59]. It can be described as a cubic close packed array of anions (O^{2-} ions) with the metal ions located in interstitial positions. Two types of interstices are available, one formed by four anions at the vertices of a tetrahedron and the other by six anions at the vertices of an octahedron; these interstices are commonly called the tetrahedral (T_d) or A sites and the octahedral (O_h) or B sites, respectively as illustrated in Figure 1.6. The ionic radii of the specific ions, size of the interstices, temperature and the orbital preference for specific coordination determine the preferences of the individual ions for the two types of lattice sites [60-63].

The positions of the ions in the spinel lattice are not perfectly regular and some distortion does occur. The T_d sites are often too small for the metal ions so that the oxygen ions move slightly to accommodate them. The oxygen ions connected with the O_h sites move in such a way as to shrink the size of the O_h cell by the same amount as T_d site expands.

Further, the spinel structure is very flexible with respect to the cations it can incorporate. There are two end-member cation distributions in the spinels, called normal and inverse spinels. Originally, in accordance with classical principles of crystallography, in the spinels of the type AB_2O_4, if the less abundant A cation is restricted to T_d sites (also called A-sites) and the more abundant B cation is restricted to the O_h sites (also called B sites) the spinels are called normal spinels. Barth and Posenjak [64] pointed out a second possibility, in which half of the B cations occupy the T_d sites and all A cations together with the other half of the B cations in the O_h sites. This type of spinel configuration is called 'inverse' spinels. In addition to these two arrangements, there are possibilities for some intermediate arrangements with an averaged distribution of all ions about all spinel cation positions. Recent work by Datta and Roy [65] and Hafner and Laves [66] have shown that there are many 'intermediate' or 'random' spinels which are in between the normal and inverse arrangements. These intermediate spinels are normally labelled in terms of the percent inverse character that they exhibit. Some of the examples of normal, inverse and intermediate spinels are presented in Table.1.1

Deviations from the ideal structure with orthorhombic and tetragonal symmetry are known, although they are scarce and poorly characterized. The distorted structure may exist at relatively low calcinations temperature, but it transforms to the cubic structure at high temperature. There are two origins for these distortions, viz. the John-Teller effect and superstructure formation.

Rao et al [67] have performed the ^{57}Fe Mössbauer and MoKα EXAFS investigation of $Mo_xFe_{3-x}O_4$. Mössbauer studies show that Fe^{+2} and Fe^{+3} ions are present on both the A and B sites, giving Fe an average oxidation states between 2+ and 3+. Molybdenum is present in +3 and +4 states on the B sites. Presence of Mo in +3 state has been established by determining the Mo^{+3}-O distance (2.2 Å), for the first time, by MoKα EXAFS. The mixed valence of Fe on both the A and B sites and of Mo on B sites is responsible for the fast electronic transfer between the cations. All the Mössbauer parameters including the line width show a marked change at composition (x=0.3), above which the concentration of the Fe^{2+} increases repeatedly.

Large and Kacim [68] has studied the cationic distribution and oxidation mechanism of trivalent Mn ions in sub micrometer $Mn_xCoFe_{2-x}O_4$ spinel ferrites. It has been suggested that valence states of Mn and Co ions in sub micrometer $Mn_xCoFe_{2-x}O_4$ spinel ferrites are +3 and +2 respectively. The Mn^{+3} ions are located in octahedral sites while Co^{+2} ions may be located in both octahedral and tetrahedral sites.

The valency state and cation distribution on the octahedral (B) and tetrahedral (A) sites in the spinel structure of Cu and Fe ions has been determine in nanometric Cu substituted Magnetite's, $Cu_xFe_{3-x}O_4$ (X=0 to 0.5) by TG, DTG, FTIR and XPS by Kester et.al [69]. The oxidation of the oxidizable cations has also been investigated. It has been demonstrated that Cu^+ and Fe^{+2} of B-sites and Cu^+ of A-sites oxidize into Cu^{2+} and Fe^{3+} ions below 300 °C.

1.6.1 *Magnetic properties of Zn-Mn ferrites*

Mn-Zn is a soft magnetic material with low Curie temperature, low remaining magnetization and low magnetic crystalline anisotropy constant. Magnetic moments of ferrites are taken as sum of magnetic moments of sub lattice A and sub lattice B. Exchange interaction between electrons of ions in these sub lattices is different and follow the order AB > AA > BB. This interaction can lead to complete or partial antiferromagnetism (ferromagnetism) [70,71].

In the inverse ferrites one half of the Fe^{3+} is placed in A-sites and another half in B-sites. Their magnetic moments are mutually compensated and the resultant magnetism is due to magnetic moment of Mn^{2+} in the B-positions.

$$\text{A} \quad [\quad \text{B} \quad]$$
$$\text{Fe}^{3+}[\text{Mn}^{2+}\text{Fe}^{2+}]\text{O}^{2-}_4$$
$$\downarrow \quad \uparrow \quad \uparrow \quad 0$$

Magnetization of $MnFe_2O_4$ with inverse degree of 0.2 can be described by following scheme.

$$\text{A} \quad\quad\quad [\quad\quad \text{B} \quad\quad]$$
$$\text{Mn}^{2+}_{0.8}\text{Fe}^{3+}_{0.2}[\text{Mn}^{2+}_{0.2}\text{Fe}^{3+}_{0.8}\text{Fe}^{3+}]\text{O}^{2-}_4$$
$$\downarrow \quad \downarrow \quad \uparrow \quad \uparrow \quad \uparrow \quad 0$$

Substitution of Mn^{2+} with nonmagnetic Zn^{2+} into A-sites leads to increase in saturation magnetization σ_s [72]. Increasing x up to 0.4 (at 0 K) leads to linear increase in magnetization and may be described by following scheme [73].

$$\text{A} \quad\quad\quad\quad [\quad\quad \text{B} \quad\quad]$$
$$\text{Zn}^{2+}_x\text{Mn}^{2+}_y\text{Fe}^{3+}_{1-x-y}[\text{Mn}^{2+}_{1-x-y}\text{Fe}^{3+}_{1+x+y}\text{Fe}^{3+}]\text{O}^{2-}_4$$
$$0 \quad \downarrow \quad \downarrow \quad \uparrow \quad \uparrow \quad \uparrow \quad 0$$

For higher Zn concentration (x > 0.4-0.5) ferrite becomes normal spinels having no Mn^{2+} in B-sites and no Fe^{3+} in A-sites. Also with x >0.4 A-B exchange interactions weaken and B-B interaction increases. Due to this, magnetic moment of Fe^{3+} ions (which are now in only B-positions) becomes oriented reversibly and leads to decrease of the total magnetic moment. Due to thermal fluctuations, the relative effect of A-B exchange weakening decreases with increase in temperature and leads to decrease in magnetization.

For x close to 1, the B-B interactions becomes dominant and in $ZnFe_2O_4$ magnetic moment of Fe^{3+} have anti-parallel orientation and the magnetic moment is zero and can be described as follows

$$\text{Zn}^{2+}[\text{Fe}^{3+}\text{Fe}^{3+}]\text{O}^{2-}_4$$
$$\uparrow \quad \downarrow$$

It can be seen that smaller size of particles yields to a lower magnetization.

1.6.2 *Catalytic activity of ferrites*

All catalytic reactions proceed through adsorption. The adsorbate atoms are trapped at sites where the well depth of the attractive surface potential is higher than the kinetic energy of the atoms. The sticking probability of atoms is defined as the rate of adsorption divided by the rate of collision, and it is higher for more open and rough surfaces than on flat ones. This is due to the higher heat of adsorption at a kink or step edge site. Decreasing size leads to surface roughing and encourages adsorption with wide range of desorption/adsorption energies [74].

Bond breaking of the adsorbate molecules is the main function of catalysts. Bonding of adsorbate to metal generally increases from left to right in the periodic table. The d-electron back bonding depends on the degree of filling of the anti-bonding states. This means that early transition metals with fewer d-electrons form stronger chemical bonds. The heat of adsorption correlates with the heats of formation of the corresponding oxides and hydrides. Molecules dissociate on more open and atomically rough surfaces at lower temperatures than on flat and close-packed surfaces.

The heat of adsorption is higher at defect sites on a surface. Defect sites, surface roughness and low packing density of surfaces give rise to higher charge densities near the Fermi level. The work function of these sites is lower and the density of filled electronic states is higher. They are likely to have more adsorbate-induced restructuring. These factors can give rise to enhanced reactivity and bond strength and lead to surface-structure sensitivity of the adsorbate bond. Atomic steps break H-H and C-H bonds efficiently while kinks in the steps are required for additional C-C and C=O bond scissions [75].

Catalytic reactions can be classified as structure insensitive and structure sensitive. For structure insensitive reactions (C-H bond formation or breaking) only one active surface metal atom is needed. For structure-sensitive reactions (C-C bond breaking/formation) several adjacent active surface atoms are needed. Particle size and other properties can effectively change catalytic activity and selectivity [76].

It has been found that in spite of their low surface area (about $40 m^2/g$) ferrospinel nanoparticle is found to be effective catalysts for alkylation and cyclization reactions. A correlation of their activity with surface area and acidity has revealed that for such materials, acidity is more important as compared to surface area. Decrease of particle size leads to increase in the Lewis acid sites responsible for their catalytic action. Besides, doping them with suitable transition metal can control acidity of such materials. Such control of acidity has pronounced effect on the yield and selectivity of reactions.

The ample diversity of properties that the spinel compounds exhibit is derived from the fact that the possibility of synthesis of multicomponent spinels by partial substitution of cations in position A and B giving rise to compounds of formula $(A_x A'_{1-x})(B_y B'_{2-y}) O_4$. This accounts for the variety of reactions in which they have been used as catalyst. Moreover, partial substitution of A and B ions giving rise to complex oxides is possible while keeping the spinel structure.

The interesting catalytic properties shown by spinels are controlled by various properties such as, nature of ions, their charge and site distribution among Td and Oh sites. Differential Neutron Diffraction (DND) studies on spinels revealed that the surface of normal spinel consists of a mixture of (110) and (111) planes.

Nanocrystalline ferrites as catalysts have been found to be very selective for the synthesis of N-alkyl aniline, alkyl pyridine and alkylated phenols [77]. Chromites and ferrites both belong to spinel family having general formula $A^{+2}[B^{+3}_2]O_4$, formed due to close packing of oxygen anions having tetrahedral and octahedral interstitial sites filled by A^{+2} and B^{+3} ions respectively. Metal ions in square bracket are in octahedral sites. Depending upon the position of metals in the tetrahedral and octahedral sites, spinels can be normal $A^{+2}[B^{+3}_2]O_4$, inverse $B^{+3}[A^{+2}-B^{+3}_2]O_4$ or mixed spinels in which the divalent cations are distributed between both sites. This type of cation distribution significantly affects magnetic, acido-basic and surface properties of spinels. Spinels are reported to be thermally stable and maintain enhanced and sustained activities for a variety of industrially important reactions [78-114].

1.7 X-Ray Diffraction Technique

1.7.1 Production of X- rays

X-rays were discovered by Conrad Roentgen in 1895. These are electromagnetic radiation with wavelengths between about 0.02 Å and 100 Å. X-rays are produced in a device called X-ray tube. It consists of an evacuated chamber with a tungsten filament as cathode and a metal target as anode. On passing electrical current the tungsten filament glows and emits electrons. A large voltage difference (kilovolts) is placed between the cathode and the anode, accelerates electrons and energizes them. The energized electron hits the target and ejects an electron from a K shell. The electrons from L and M shells jump to K shell to replace the ejected electron. These electronic transitions results in the generation of X-rays (Fig. 1.7). A transition from the L - shell to the K- shell produces a K_α X-ray, while the transition from an M - shell to the K- shell produces a K_β X-ray. A filter is generally used to filter out the lower intensity K_β X-rays. Some commonly used target materials and their K_α lines are listed in Table 1.2

In recent years, synchrotron facilities have become widely used as preferred sources for x-ray diffraction measurements. Synchrotron radiation is emitted by electrons or positrons traveling at near light speed in a circular storage ring. These powerful sources are thousands to millions of times more intense than laboratory x-ray tubes.

1.7.2 X-ray diffraction by crystals

Crystals consists of atoms, ions, or molecules arranged in a periodic array. The inter-planer distances in the crystals matches with the wavelength of X-rays which causes diffraction of X-rays by crystals according to the Bragg's law (Fig. 1.8).

A graph between 2θ and intensity of diffracted light is called diffraction pattern (XRD Pattern) and is characteristic of a crystal. A typical XRD pattern for $MnFe_2O_4$ is shown in Fig. 1.9. Thus, every crystalline materials can be recognised by comparing the observed pattern with standard pattern of the material available in literature (say JCPDS cards/ ASTM cards). X-ray diffraction (XRD) analysis of materials is unique because, it discloses the presence of a particular material and not the elemental or functional groups present in the materials. This is the basic difference in between molecular spectroscopy and X-ray diffraction.

1.7.3 The powder method

In order to apply Bragg's law to determine inter-planer distances of a single crystal, we need to re-orient the crystal many a times to satisfy Bragg's law condition for different planes. This is a time consuming operation to reorient the crystal, measure the angle θ, and determine the d-spacing for all atomic planes. 0Powder method provides a faster technique to record XRD patter. In this method, a powder sample is used in place of single crystal. With random orientations of crystals there will be some atomic planes which will satisfy the Bragg condition and will produce constructive interference. Thus, by scanning through an angle θ of incident X-ray beams form 0 to 90°, we would expect to find all angles where diffraction has occurred, and each of these angles would be associated with a different atomic spacing. The experimental geometry for powder diffraction is explained in chapter 2.

1.7.4 Confirmation of phase purity from XRD Pattern

Phase purity of a sample is confirmed by matching the observed pattern with its standard pattern available in literature. If both patterns are same the material is pure. Appearance of extra peaks indicates the presence of impurity material. A Comparison of inter planar distances, d and corresponding intensities for standard $ZnFe_2O_4$ and $MnFe_2O_4$ with synthesized $ZnFe_2O_4$ and $MnFe_2O_4$ is presented in Table 1.3 Excellent agreement between observed pattern and standard ASTM pattern confirms the phase purity of the samples.

1.7.5 Particle size from XRD

The mean diameters of particles are determined from the measurements of the widening at half height through the Scherrer formula [112]:

$$D_{XR} = \frac{0.9 \times \lambda}{\Delta\theta \times \cos\theta} \qquad \ldots\ldots\ldots\ 1.7.1$$

1.7.6 Determination of unit cell type and indexing of planes

A further confirmation of the unit cell type and quantitative treatment of observed pattern was made by combing Bragg law with plane-spacing equation for cubic system. The perpendicular distance, d, between adjacent hkl planes for an orthorhombic unit cell is given by

$$1/d^2 = h^2/a^2 + k^2/b^2 + l^2/c^2 \qquad \ldots\ldots\ldots\ 1.7.2$$

which reduces to

$$1/d^2 = \frac{h^2 + k^2 + l^2}{a^2} \qquad \ldots\ldots\ldots\ 1.7.3$$

for cubic lattice. Combination of this equation with Bragg equation $n\lambda = 2d\sin\theta$ yield

$$\sin^2\theta = \frac{\lambda^2 (h^2 + k^2 + l^2)}{4a^2} \qquad \ldots\ldots\ldots\ 1.7.4$$

$$\frac{\sin^2\theta}{(h^2 + k^2 + l^2)} = \frac{\sin^2\theta}{S} = \frac{\lambda^2}{4a^2} \qquad \ldots\ldots\ldots\ 1.7.5$$

This equation can be used to determine lattice type and lattice plane.

The sum $S = (h^2 + k^2 + l^2)$ is always integral and can have only following set of values for different cubic unit cells.

- Simple cubic: 1, 2, 3, 4, 5, 6, .8, 9, 11, 12, 13, 14, 16 ...
- Body centered cubic: 2, 4, 6, 8, 10, 12, 14, 16 ......
- Face centered cubic: 3, 4, 8, 11, 12, 16.........

Certain integral values of S such as 7, 15, 23, 28, 31 etc are not possible because they cannot be formed by the sum of three squared integers. Since $\lambda^2 / 4a^2$ is a constant, the problem of indexing cubic pattern is simply to divide the observed $\sin^2\theta$ values successibly by the set of S values for simple cubic, body centered cubic and face centered cubic systems and to observe, which set produces a constant real number for all values of $\sin^2\theta$. The unit cell that corresponds to this set is the unit cell of the substance.

1.7.7 Intensity of X-ray diffraction

Diffraction involves simultaneous scattering and interference. The scattering results due to interaction between the X rays and the sample. The scattered radiation beams interfere with one another because of the periodic nature of the crystalline sample. While scattering can be observed in all kind of materials, diffraction is possible only with crystals possessing ordered periodic structure. The elastic X-rays scattering occurs due to oscillation of induced dipole in the atom produced due to incident X-ray photons. It is worth mentioning here that Raman scattering is inelastic.

X-rays are scattered by the electrons in crystals. The function describing the scattering of X-rays by an atom is called the atomic scattering factor. It depends upon electron density (number of electrons per unit volume), $\rho(r)$, as well as angle of diffraction and is expressed by.

$$f = 4\pi \int \rho(r) \frac{\sin kr}{kr} r^2 dr \qquad \ldots\ldots 1.7.6$$

Here $k = (4\pi/\lambda) \sin \theta$, where θ is the scattering angle and λ is the wave length of X-ray. When $\theta = 0$, $k = 0$ and $(\sin kr)/kr$ becomes indeterminant and it value approximates to be 1 as $kr \to 0$ and

$$f = 4\pi \int \rho(r) r^2 dr \qquad \ldots\ldots 1.7.7$$

Equation 2 after integration, is simply number of electrons in the atom. Thus atomic scattering factor is simply equal to atomic number of the atom when glancing angle is zero. Atomic scattering factors of different atoms are available in literature.

It is assumed that the electron density of each atom is a discrete and spherically symmetric entity. However this could be a poor approximation for heavy atoms with considerable amounts of d type electrons. The scattering functions are independent of the wavelength of radiation and only depend on the scattering angle and number of electrons.

It is also obvious from equation that f will decrease with increase in θ. This happens because X-ray photons hitting different parts of the electron cloud of an atom are not expected to scatter in phase with one another. Also, the more diffuse the electron cloud, the more rapid will be the reduction in the scattering function with scattering angle. For example both Ca^{2+} and Cl^- have 18 electrons. Each of these are expected to have same value of f at zero scattering angle. However, at higher scattering angle, the Cl^- species would be expected to have a smaller f than Ca^{2+} and because it has a more diffuse electron cloud.

The structure factor of a plane, F(hkl) with miller indices hkl is given by summing the scattering from all the atoms of the unit cell and it can be expreesed as,

$$F(hkl) = \Sigma f_j e^{2\pi i (hx_j/a + ky_j/b + lz_j/c)} \qquad \ldots\ldots 1.7.8$$

where a, b and c are the lengths of the three sides of the unit cell, f_j is scattering factor of an atom of jth type having cartesian coordinates x_j, y_j and z_j. The above equation can also be written as

$$F(hkl) = \Sigma f_j e^{2\pi i (hx_j' + ky_j' + lz_j')} \ldots \qquad \ldots\ldots 1.7.9$$

Intensity of a spot from a plane of the crystal is proportional to the square of the magnitude of the structure factor, i.e.,

$$I \ \alpha \ |F(hkl)|^2 \qquad \ldots\ldots 1.7.10$$

Thus, only those set of planes will produce diffraction spots for which F(hkl) is non zero. A simple consideration will show that for a cubic unit cell of identical atoms all lattice planes of simple cubic cell with positive values of h, k and l will produce reflections, while planes possessing either all even hkl or all odd hkl values of a face – centered cubic unit cell will produce reflections and in case of body centered cubic unit cell only planes with h + k +l a even number will produce reflections. Above discussion shows it is possible to identify the type of unit cell from the consideration of intensity of diffracted rays.

More rigorously, electron density is considered to be a continuous function of coordinates and therefore Σ in equation 1.7.9 can be replaced by $\int$. Further, f can be replaced by $\rho(x,y,z)$ and equation can be written as

$$F(hkl) = \int_{Cell} \rho(x,y,z) \ e^{2\pi i (hx_j' + ky_j' + lz_j')} \ dxdydz \qquad \ldots\ldots 1.7.11$$

The Miller indices (h k l), besides representing the Bragg planes, also represent a vector in reciprocal space perpendicular to Bragg planes, with length equal to the reciprocal of the spacing between the planes. The amplitude |F(h k l)| of a particular structure factor indicates the extent to which the electron density is concentrated on planes parallel to the Bragg planes, while its phase indicates the position of planes of high electron density relative to the Bragg planes.

One of the most useful properties of the Fourier transform is that it is its own inverse: Thus if the Fourier transform is applied twice, the original function is obtained back. In the inverse Fourier transform, the only difference will be that the a negative sign will appear in the exponential, and (depending exactly on how the first Fourier transform was defined), there may be a scale constant.

$$\rho(x,y,z) = \sum_{hkl} F(hkl)\, e^{-2\pi i (hxj' + kyj' + lzj')}\, dxdydz \qquad \ldots\ldots 1.7.12$$

So we can regenerate the electron density from the structure factors with an inverse Fourier transform. The equation is called the electron density equation. In general, an inverse Fourier transform would involve an integral like the forward Fourier transform, but if the object is periodic (like a crystal), it involves just a summation. (because diffraction from a crystal cancels out in all directions, except those specified by integer Miller indices.)

Since $F(hkl)$ is a complex quantity, it can be written as

$F(hkl) = A(hkl) + iB(hkl)$

Then intensity equation 1.7.10 can be written as

$I(hkl) \alpha \ |F(hkl)|^2 = [A(hkl) + iB(hkl)][A(hkl) - iB(hkl)]$
$\qquad\qquad = |A(hkl)|^2 + [B(hkl)]^2$

Unfortunately, $A(hkl)$ and $B(hkl)$ can not be determined separately from XRD experiments, but only sum of their square can be measured. The methodology of determining $A(hkl)$ and $B(hkl)$ from $I(hkl)$ is called phase problem. Crystallographers have developed methods to slove this problem.

Once the phase problem is solved, the electron density map of a molecule can be constructed from the knowledge of x-ray diffraction pattern.. The position of nuclei can be deduced from such maps, leading to the knowledge of bond length and bond angles. An electron density map of molecule is shown in Fig.1.10.

1.8 Mössbauer Spectroscopy

Mössbauer spectroscopy involves nuclear transitions that result from the absorption of γ-rays by the sample. The decay of a gaseous γ-ray source can be written as

$$E_\gamma = E_r + D - R \qquad \ldots\ldots 1.8.1$$

Where E_r is the difference in energy between the exited state and ground state of the source nucleus, D is the Doppler shift due to translational motion of the nucleus, and R is the recoil energy of the nucleus [115], similar to that found when bullet leaves a gun and is given by equation

$$R = E_r^2 / 2mc^2 \qquad \ldots\ldots 1.8.2$$

It is of the order 10^{-2} to 10^{-3} eV.

The energy of the gamma ray absorbed for a transition in the sample is given as

$$E_\gamma = E_r + D + R \qquad \ldots\ldots 1.8.3$$

Here R is added because the exciting gamma rays have energy necessary to bring about the transition and effect recoil of the absorbing nucleus. E_r is taken to be same for source and sample.

It is easy to understand that the major difference in source and sample energies is due to R and the matching between source and sample energy is possible by minimizing R, which is possible by taking the sample as well as the source in solid phase. Under this condition, it is possible to match the energies of source and sample by a change in Doppler effect by introducing a velocity of 1 mm/s.

1.8.1 Isomer Shift

Different line positions are obtained for iron atoms present in different chemical environment. The shift in line position measured with respect to metallic iron is known as isomer shift. Isomer shift results due to electrostatic interaction of the charge distribution in the nucleus with s-electron density $\Phi_s^2(0)$.

In Mössbauer spectroscopy source and sample both contain iron nucleus, therefore R can be taken as constant for both. However, $\Phi_s^2(0)$ differs for the two because of different chemical environment.

Electron in the p or d- orbitals can change the electron density of the s-orbital by screening the s-density from the nuclear charge. It has been shown that decrease in number of d-electrons increases the total $\Phi_s^2(0)$ at the iron nucleus. Isomer shift also depends on the type of ligands attached to the iron atom. Correct conclusion can be drawn only if ions are examined in series of molecules.

1.8.2 Electric Quadrupole Interaction

When charge distribution around a nucleus is non spherical the nucleus is said to possess a quadrupole moment eQ, where Q is the measure of deviation of the nuclear charge distribution from spherical symmetry. When Q is +ve, nuclear orientation is along principal axis (Pl) and when Q is –ve, the orientation is perpendicular to the Pl.

1.8.3 Magnetic Splitting

The magnetic splitting arising due to removal of degeneracy of $\pm 1/2$ and $\pm 3/2$ states in presence of magnetic field. In presence of external magnetic field or due to internal magnetic field of a ferromagnetic sample, the degeneracy of $\pm 1/2$ and $\pm 3/2$ states is removed and a symmetric six line spectrum for ^{57}Fe is achieved. In case of diamagnetic materials, the two line zero field spectrum splits into a doublet or triplet or for a small η, the doublet arises from $+1/2 \rightarrow +3/2$ and $-1/2 \rightarrow +3/2$ transitions. If this doublet lies toward positive velocity for ^{57}Fe, the signs of the quadrupole splitting and q are positive.

Low value of quadrupole splitting (Q.S.) in case of $ZnFe_2O_4$ (0.48 mm s^{-1}) is typical of Fe^{3+} in octahedral surrounding. The increase in Q.S. to 0.66 mm s^{-1} for $CuFe_2O_4$ seems to be caused by Jahn Teller effect due to copper ions which prefer octahedral sites. The Q.S. is maximum (0.79 mm s^{-1}) in case of mixed ferrite i.e. $Zn_{0.5}Cu_{0.5}Fe_2O_4$ which may be attributed to the presence of two different cations instead of one introducing a kind of lattice stress and introduce additional asymmetry.

1.9. Surface area of catalyst by Brunauer Emmett and Teller (BET) Method

Theory: Experimentally it has been found that adsorption isotherms of gases near their condensations points, for most adsorbents, are concave at low pressure and convex at higher pressure towards the pressure axis (S shaped) Fig 1.11.

The high pressure convex portion has been attributed to condensation in capillaries of the adsorbent and to the formation of multimolecular adsorbed layers. The BET equation [116] is generalization of the Langmuir's theory of monomolecular adsorption of ideal gases to multimolecular adsorption including vapours also. The equation is capable of (1) representing general shape of the experimental curves (2) yielding reasonable value for the average heat of adsorption in the first layer and (3) evaluating volume of the gas required for unimolecular adsorption.

In BET method a powdered sample is taken in a sample tube and cooled down to temperature at which nitrogen liquefies at atmospheric pressure. Nitrogen gas of known volume, at about 10-20 mm of pressure, is allowed to enter the sample tube where some amount of gas is adsorbed by the powder. The methodology allows determination of volume of nitrogen necessary to form a monolayer over the catalyst. This volume is used to calculate the number of nitrogen molecules which when multiplied with surface area of 1 nitrogen molecule ($12\ A°^2$) gives the surface area of the catalyst.

Let $\theta_0, \theta_1, \ldots \theta_n$ be the surface area covered by 0, 1st... nth layer of the adsorbed molecules. Since at equilibrium θ_0 must remain constant:

Rate of condensation onto bare surface = Rate of evaporation from first layer

$$k_1 P\, \theta_0 = k_{-1}\, \theta_1 e^{-E_1/RT} \qquad \ldots\ldots 1.9.1$$

Here P is pressure of the gas, E_1 is heat of adsorption of the fist layer and k_1 and k_{-1} are adsorption equilibrium constants. It is worth mentioning that (1) rate of condensation is proportional to pressure of the gas and available bare surface (2) rate of evaporation is proportional to covered surface and heat of adsorption and at equilibrium rate of adsorption is equal to rate of desorption.

Similarly at equilibrium θ_1 must remain constant, therefore

Rate of condensation on the bare surface + rate of evaporation from the second surface =

Rate of condensation on the first layer + rate of evaporation from the second layer

$$k_1 P\, \theta_0 + k_{-2}\, \theta_2\, e^{-E_2/RT} = k_2 P\, \theta_1 + k_{-1}\, \theta_1\, e^{-E_1/RT} \qquad \ldots\ldots 1.9.2$$

With the help of 1.9.1, we get

$$k_{-2}\, \theta_2\, e^{-E_2/RT} = k_2 P\, \theta_1 \qquad \ldots\ldots 1.9.3$$

This can also be interpreted as the rate of condensation on the top of first layer is equal to the rate of evaporation from the second layer. Generalizing this argument to other layers

$$k_{-i}\, \theta_i\, e^{-E_i/RT} = k_i P\, \theta_{i-1} \qquad \ldots\ldots 1.9.4$$

The total surface area of the catalyst is given by $A = \sum_{i=0}^{\infty} \theta_i \qquad \ldots\ldots 1.9.5$

Total volume of the gas adsorbed on the surface, $V = V_0 \sum_{i=0}^{\infty} i\theta_i$ 1.9.6

Where V_0 is the volume of gas adsorbed on one cm² of surface when it is covered with a complete layer. Therefore,

$$\frac{V}{AV_0} = \frac{V}{V_m} = \frac{\sum_{i=0}^{\infty} i\theta_i}{\sum_{i=0}^{\infty} \theta_i}$$ 1.9.7

Where V_m is the volume adsorbed when the entire surface is covered with a complete monolayer.
The summation in equation 1.9.6 is possible if we make following assumptions.
1. $E_2 = E_3 = \ldots E_i = E_L$ where E_L = heat of liquefaction and
2. $k_{-2}/k_2 = k_{-3}/k_3 = \ldots = k_{-i}/k_i = K$

Condition 2 is based on the assumption that properties of 1st, 2nd ……. layers are equivalent
Now we can express $\theta_1, \theta_2, \theta_3 \ldots$ In terms of θ_0

From 1.9.1 $\theta_1 = \{(k_1/k_{-1})P\, e^{E_1/RT}\}\theta_0 = y\,\theta_0 \quad \{y = (k_1/k_{-1})P\, e^{E_1/RT}\}$ 1.9.8

$\theta_2 = x\,\theta_1 \quad\quad\quad \{x = p/K\, e^{E_1/RT}\}$ 1.9.9

$\theta_3 = x\,\theta_2 = x^2\,\theta_1$ 1.9.10

$\theta_i = x\,\theta_{i-1} = x^{i-1}\,\theta_1 = y\,x^{i-1}\,\theta_0 = c\,x^i\,\theta_0 \quad \{c = y/x\}$ 1.9.11

Substituting into 1.9.7 we get

$$\frac{V}{AV_0} = \frac{V}{V_m} = \frac{c\,\theta_0 \sum_{i=0}^{\infty} i x^i}{\theta_0\{1 + c \sum_{i=0}^{\infty} x^i\}}$$ 1.9.12

Now $\sum_{i=0}^{\infty} x^i = x/(1-x)$ { sum of an infinite geometrical series } 1.9.13

and

$\sum_{i=0}^{\infty} i x^i = x(d/dx) \sum_{i=0}^{\infty} x^i = x/(1-x)^2$ 1.9.14

Substituting the values of x^i and ix^i in equation 1.9.12 we get,

$$\frac{V}{V_m} = \frac{cx}{(1-x)(1-x+cx)}$$ 1.9.15

When $p = p_0$, i.e. saturation vapor pressure of the gas, p_0 an infinite numbers of layers can be formed on the catalyst surface and $v = \infty$. For this condition to be satisfied x must be 1 in 1.9.15 and from eqn. 1.9.9

$\{x = p/K\, e^{E_1/RT}\} = 1; \; K = p_0 \text{ and } x = p/p_0$ 1.9.16

Substituting the value of x in equation 1.9.15 we obtain the BET isotherm

$$V = \frac{V_m cx}{(1-x)(1-x+cx)} = \frac{V_m cp/p_0}{(1-p/p_0)(1-p/p_0+c\,p/p_0)} = \frac{V_m cp}{(p_0-p)\{1+(c-1)p/p_0\}}$$ 1.9.17

Plot of equation 1.9.17, is a S shaped curve as found experimentally. Following form of equation is useful for testing

$$\frac{P}{V(p_0-p)} = \frac{1}{V_m c} + \frac{(c-1)}{V_m c}\frac{p}{p_0} \quad\quad\quad ....1.9.18$$

A plot of $p/\{V(p_0-p)\}$ vs p/p_0 will result into a straight line having intercept = $1/V_m c$ and slop = $c-1/V_m c$ permitting evaluation of two constants V_m and c. From the knowledge of V_m surface area can be calculated.

In a typical experiment about 0.1- 0.5g of the sample is taken into a spherical Pyrex-glass bulb attached to the system through a stopcock. The sample is degassed at 150 ^0C. A vacuum of 10^{-6} torr is maintained for a considerable time to ensure that no pre-adsorbed gas is present on the catalyst surface. Dead space calibration has to be carried out at liquid nitrogen temperature by using helium gas in successive steps by measuring the volume at different pressures. After degassing of helium gas upto a 10^{-6} torr vacuum, the nitrogen adsorption is carried out at liquid nitrogen temperature in the same way as performed for the dead space determination with helium. The adsorbed nitrogen volume (V) is obtained from the difference between the volume of nitrogen introduced and the volume of the helium introduced into the sample tube at the equilibrium pressure. From the measured value of P_o and V the specific surface area of the catalyst sample was calculated.

1.10 *Gas Chromatography*

Gas Chromatography is a separation technique in which the solute mixture to be separated is distributed between two phases, a mobile phase, generally, dry nitrogen and a stationary phase (a liquid phase immoblized on the surface of an inert solid such as a diatomaceous earth on the walls of a capillary tube). The technique involves a sample being injected and vaporised onto the head of the chromatographic column. The sample is then transported through the column by the flow of inert, gaseous mobile phase. The vaporized sample of the mixture to be separated is partitioned or adsorbed on the column depending on the partition coefficient of the components of the mixture between the selected liquid phase and the mobile phase.

Partition coefficient (K) may be defined as:

$$K = \frac{\text{Concentration of solute in the stationary phase}}{\text{Concentration of the solute in the mobile phase}}$$

A schematic diagram of the apparatus is shown in Figure.1.12. The apparatus basically consists of following components.

1. Flow controllers 2. Injector port 3. Column 4. Oven 6. Detector 7. Recorder

1. *Flow Controller:* Generally nitrogen, hydrogen and air are needed for running a GC experiment. Nitrogen is used as a carrier gas. Hydrogen along with air is needed for burning the organic compound in the flam ionization detector. Flow rates are normally controlled by a two stage pressure regulator, one at the gas cylinder and another mounted in the chromatograph.

2. *Injector port:* The sample is injected into the hot injector where it is vaporized. The carrier gas (in most cases: He or N_2) transports the vaporized sample through the thermostatic capillary column into the detector, where the separated analytes are obtained as peaks in a chromatogram. The primary goal for any injection is to introduce the sample into the column. Liquid samples can be injected with small micro syringes (volume 1–10μl for capillary columns and 20μl for packed columns). Gaseous samples can be injected with gas-tight glass-syringes with volumes between 1 and 10ml. A better way is injecting gaseous samples with a sample loop of a defined volume. In all cases, the injection has to be reproducible without disturbing the equilibrium state in the column. The sample has to be vaporized in a small zone, and must produce narrow peaks for every compound. Two types of injectors, split and splitless are commonly used in Gas Chromatography. Each type of injector is well suited for a particular type of sample. A schematic diagram of a split/splitless injector is shown in Figure 1.13.

It consists of a glass tube ("liner") which creates an inert environment inside the injector and where the sample is vaporized. A syringe is used to pierce the septa and to introduce the sample into the injector. The injector temperature should be high enough to ensure instant vaporization without degradation of the sample. The injector port is ordinarily about 50^0C above the boiling point temperature of the least volatile component of the sample. The sample vapour is mixed with the carrier gas and is transported into the column. Components of the sample that are not vaporized remain in the injector. The septum purge is a low flow, which minimizes the amount of septum bleed materials, which could contaminate the GC system. Septum purge gas sweeps the bottom of the septum and the top of the liner, out through the purge vent. A typical septum purge flow is between 0.5 and 5 mL/min.

3. *Column:* The separation of components takes place in the column by physico-chemical interaction of the sample compounds with the stationary and the mobile phases. The separation depends on stationary phase type, stationary film thickness and column inner diameter (factors which are interrelated), and column length. Depending upon the analyte one can either use packed column (for fine chemicals) or the capillary column (for lipids). The columns are made up of stainless steel / glass and are from less than 2 m to 50 m in length. Columns are generally coiled to accommodate them into the oven. the columns are packed with stationary liquid phase materials on solid support. the most widely used solid support is naturally occurring diatomaceous earth and most commonly used liquid stationary phases are polydimethyl siloxanes having general formula

$$R-\underset{R}{\overset{R}{\underset{|}{\overset{|}{Si}}}}-O-\left[\underset{R}{\overset{R}{\underset{|}{\overset{|}{Si}}}}-O\right]_n-\underset{R}{\overset{R}{\underset{|}{\overset{|}{Si}}}}-R$$

The selectivity is influenced by polarizability, solubility, magnitude of dipoles and hydrogen bonding behavior of the stationary phase with the different compounds. If the stationary phase retains one compound to a greater extent than another, the compounds can be separated. The selection of the stationary phase is based on the following principle 'a non-polar column is better for separation of non-polar analytes', and polar columns effectively separate polar analytes.

The retention time (t_R) is directly influenced by the column length. Longer columns cause higher retention times. Capillary column length ranges between 10m up to 100m, but commonly 25m or 60m columns are used.

The column temperature influences the separation, too. An increased temperature induces a decrease of the distribution coefficient and the retention time. Simultaneously the viscosity of the carrier gas increases, which leads to an increasing dead time and a decreasing gas flow.

The mechanism involved in separating different components present in a mixture employing chromatographic techniques is based on plate theory applied in distillation columns. The column efficiency can be measured, either by stating the number of theoretical plates in a column, N (the more plates the better), or by stating the plate height; the Height Equivalent to a Theoretical Plate (the smaller the better).

If the length of the column is L, then the Height Equivalent to a Theoretical Plate (HETP) is $HETP = L/N$

The number of theoretical plates that a real column possesses can be found by examining a chromatographic peak after elution. The term is a measure of the efficiency of a gas chromatographic column. The efficiency of the column is better if the value of N is high.

4. *Oven:* The oven is designed to accommodate packed micro-packed and capillary columns. They have a capacity of more than 15 litres and are normally made of stainless steel and is circular in shape. Special blower is used for air circulation, so as to ensure a uniform temperature distribution. Provision are provided to cool down the oven in a short span of time by supplying cool air to the internal parts and sending the hot air out through the exhaust vent provided. There are safety cutouts provided, so as cut off the main power supply in case of depletion of carrier gas flow.

5. *Detector:* The detector converts analytes into an electrical signal which should be linear depending on the amount or concentration of the analytes. All measurements must be done in the linear range of the detector. The quantification of the compounds could be done by a multi-point calibration. For all detectors, one or more gases are necessary like combustion, reagent, auxiliary and make-up gases.

The Flame ionization detector (FID) indicates compounds with C-H-bonding. The sample is burned in an H_2/air-flame and the produced ions change the electrical conductivity. The electrical conductivity of the H_2/air-flame is low due to low ionization. Organic molecules usually burn with formation of ions and liberation of electrons, thus increasing the electrical conductivity of the flame.

The Electron capture detector (ECD) indicates electronegative compounds, which catch electrons (emitted from ^{63}Ni) and thereby reduce the current. The ECD is more sensitive than the FID, so it is used for samples with low analyte concentrations.

Other types of detector being used are TCD (Thermal Conductivity detector), PID (photo ionization Detector), MS (mass selective detector), and NPD (Nitrogen phosphorous detector).

6. *Recorder:* The chromatographs are recorded using an online electronic integrator, which integrates the peak and prints them. Nowadays, modern GC is equipped with computers with advanced data acquisition software, which acquires data and stores them.

1.10.1 Qualitative analysis

Qualitative identification of different components of a mixture in gas chromatography is carried out by comparison of retention data from known and unknown samples. The data may be expressed in terms of retention times or retention volumes. Adjusted retention time (t'_R) is defined as the solute retention time minus the retention time for an unretained peak like air and is given by

$$t'_R = t_R - t_m \qquad \ldots\ldots\ldots\ 1.10.1$$

Retention volume is obtained by multiplying the retention time by the carrier gas flow rate. Adjusted retention volume (V'_R) is calculated using the following equation,

$$V'_R = V_R - V_M \qquad \ldots\ldots\ldots\ 1.10.2$$

1.10.2 *Quantitative Analysis*

GLC can also be used for quantitative analysis. The peak size obtained is proportional to the amount of material contributing to that peak provided the detector response is linear. Measurement of the peak height and peak area is the two general methods used to estimate the size of the peak. Different methods are employed for the measurements of peak area. These are height, width at half height, triangulation, cut and weigh, disc integrator and the latest technique are electronic integrators and computers.

A number of standardization methods are available for quantitative analysis in gas chromatography. These can be classified as (i) triangulation (ii) internal normalization (iii) external standardization and (iv) internal standardization. Internal standardization method is briefly described in the following lines

Standard mixture containing known weight of the chosen internal; standard and the analyte are made and GLC analysis is carried out. Response factor is calculated, which is the ratio of the signal to sample size. When signal is peak height, the sample size is the mass flow rate through the detector at the peak maximum. A graph is plotted with the area ratios of the analyte and the standard is abscissa and the weight ratio is ordinate. It would be linear for a given system. To estimate the amount of the analyte in unknown sample, a known weight of the internal standard is added to a known weight of the sample. From the chromatogram of this mixture, the area ratio of the internal standard to the analyte is calculated. Knowing the area ratio of the analyte and the response factor, the amount of the analyte in the sample is worked out using the formula

$$W_x/W_s = A_x/A_s \times F \qquad \ldots\ldots\ldots\ 1.10.3$$

The percentage of unknown compound in the sample mixture is determined by

$$\%X = \frac{W_x}{\text{Sample weight}} \times 100 \qquad \ldots\ldots\ldots\ 1.10.4$$

1.11 *Recent Literature Survey*

Xue Dong et. al[117] have found that Cu/SiO_2 catalyst prepared by incipient wetness method exhibited very high activity and selectivity for the vapour-phase synthesis of N-butylaniline from aniline and 1-butanol, When Cu loading was 0.70 mmol/g-SiO_2 and the catalyst precursor was calcined at 500 °C, 1-butanol conversion reached 99%, and the selectivity of N-butylaniline exceeded 97%.

Catalytic vapor phase alkylation of aniline with ethanol was studied by Bankuplli et al. [118] in the presence of a zeolite catalyst in a fixed bed reactor at atmospheric pressure and moderate temperatures. Experimental runs were carried out at five different temperatures (350, 370, 390, 410 and 430 °C) with a constant aniline to ethanol ratio (1: 5) and different catalyst loadings to find the limits of external and internal resistance zones. The resistance due to external mass transfer was overcome by operating at high velocities and the pore diffusion resistance was nullified by conducting experiments with different particle sizes of the catalyst. The effectiveness factor was determined experimentally and the optimum zone of operation where the diffusional resistances are absent was obtained for kinetic experiments to be conducted. The results indicated that below 0.6 cm/sec velocity external mass transfer was likely to influence the rate of reaction. The effect of pore diffusion was accounted for by the effectiveness factor which was nearly unity for an operative range of particle size < 2 mm. Pore diffusion will therefore playa role for a particle size > 2 mm.

Various compositions of chromium manganese ferrospinels catalyts were tested by Nishamol et al [119], for the vapour phase alkylation of aniline with methanol. The samples were prepared by room temperature co-precipitation technique and characterized by various physico-chemical methods. The acidity-basicity determination revealed that the samples possess greater amount of basic sites than acidic sites. All the ferrite samples proved to be selective and active for N-monoalkylation of aniline leading to N-methyl aniline. $Cr_{0.6}Mn_{0.4}Fe_2O_4$, $Cr_{0.8}Mn_{0.2}Fe_2O_4$ and $CrFe_2O_4$ exhibited cent percent selectivity for N-methyl aniline. Neither C-alkylated products nor any other side products were detected for all catalyst samples. The catalytic activity of the samples studied in this reaction is related to their acid-base properties and also on the cation distribution. Under the optimised reaction conditions all the systems showed constant activity for a long duration.

The reaction of aniline with ethanol was carried out by Frank et al[120] over an industrial niobic acid catalyst in a fixed bed reactor at atmospheric pressure and 220-260 °C. The main products, N-ethylaniline and N,N-diethylaniline were formed consecutively. A kinetic study including a model discrimination between several Hougen/Watson type rate equations led to an Eley/Rideal mechanism, where the reaction of gas phase aniline with adsorbed ethanol is the rate determining step. As second adsorbing agent, water inhibits the reaction in higher partial pressures. N-alkylation was the main reaction observed but the addition of water decreased the selectivity and up to 15% C-alkylated products were found. The apparent activation energies for the first and second N-ethylation are 85.6 and 70.7 kJ/mol, respectively. The high equilibrium constants indicate a nearly irreversible reaction.

Yuvraj and Palanichamy [121] invastigated Ethylation of aniline over various alkali and alkaline earth metal exchanged zeolites-Y. The basic zeolites-Y were found to have higher activity than acidic zeolites-Y for N-alkylation.

Vapour phase alkylation of aniline with ethanol is studied by Narayan et al [122] over a series of H-ZSM-5 zeolites containing SiO_2/Al_2O_3 ratios of 30, 50, 70, 150 and 280. The influence of feed rate, temperature and pressure on activity and selectivity of the products, namely N-ethylaniline and N,N'-diethylaniline are investigated. The effect of acidityand SiO_2/Al_2O_3 ratio of H-ZSM-5 on alkylation activity is discussed. H-ZSM-5 with a SiO_2/Al_2O_3 ratio of 70 seems to have the acid sites required for optimum aniline alkylation activity and selectivity of the products.

A promising process for the industrial synthesis of anisole via the vapor-phase alkylation of phenol with methanol in the presence of commercial NaX zeolite was developed by Kirichenko et al [123]. A kinetic model of phenol alkylation with methanol was derived.

Charry et al[124] prepared a series of vanadium oxide supported on zirconia catalysts with varying vanadium oxide content (2.5- 12.5 wt.%) by wet impregnation method. The catalysts were characterised by X-ray diffraction (XRD), temperature-programmed reduction (TPR) of hydrogen, temperature- programmed desorption (TPD) of NH_3 and specific surface area measurements. X-ray diffraction patterns indicate the presence of crystalline V_2O_5 phase from 7.5 wt.% of V_2O_5 on zirconia. TPR patterns reveal the presence of V^{3+} species on zirconia. Ammonia TPD results suggest that acidity of the catalysts increased with increase in V_2O_5 loading up to 7.5 wt.% and decreased with further increase in V_2O_5 loading. The catalytic properties were evaluated for the vapour-phase alkylation of phenol with methanol. During alkylation of phenol with methanol exclusively C-alkylated (alkylation proceeds through ring) products were formed. The selectivity towards 2,6-xylenol was high when compared to 0-cresol. The activity of the catalysts was found to increase with V_2O_5 loading up to 7.5 wt.% of V_2O_5 and decreased with increase in further vanadia content.

Sarla Devi et al [125] investigated vapour phase synthesis of anisole by O-alkylation of phenol with methanol over lanthanum, cerium, samarium, and antimony phosphate catalysts promoted with cesium hydroxide. Among various catalysts investigated, the cesium promoted samarium phosphate provided better activity and selectivity. The effect of temperature, contact time, time-on-stream, reusability, and up-scaling of the catalyst were also studied. These studies clearly reveal that the Cs-Sm combination is the superior catalyst for selective O-alkylation of phenol with methanol. The unpromoted catalysts provided more C-alkylated side products. Incorporation of cesium suppressed the formation of side products. The X-ray diffraction analysis of various samples revealed that there is no change in the crystalline composition of the catalysts up on addition of cesium promoter. However, the surface acidity of the catalyst was observed to decrease after the incorporation of cesium promoter as revealed by the temperature programmed desorption study of anhydrous ammonia.

Vapour phase selective O-alkylation of phenol with methanol, ethanol, n-propanol and n-butanol has been investigated by Bal Rajaram and Sivasanker[126] in the temperature range 573-673 K over alkali loaded (Li, Na. K and Cs) fumed silica. Fumed silica has negligible alkylation activity, but on impregnation with alkali metal oxides (Li, Na, K and Cs) becomes active for O-alkylation producing arylalkyl ethers. Activities of the catalysts increase with metal loading and with basicity of the metal ions (Cs > K > Na > Li). Very high conversion (90%) and 100% 0-methylation selectivity was obtained over Cs loaded silica with methanol as the alkylating agent. Conversion decreases with increase in carbon chain of the alkylating agent. Catalyst deactivation rate and the reaction rates at various reaction parameters like contact time, temperature, mole ratio, were also investigated.

Vapour phase alkylation of phenol with methanol over magnesium-aluminium calcined hydrotalcites (MgAl-CHT) with an Mg/ Al atomic ratio of 3-10 has been investigated in detail and reported previously. This study concerns the evaluation of the kinetic parameters such as the activation energy (Ea) and Arrhenius frequency factor On A_0) for the disappearance of phenol employing the Power law equation assuming pseudo first order kinetics. The kinetic parameters were found to be in good agreement with the specific activity of the catalysts. The existence of the compensation effect between E_a and ln A_0 has been tested by Velu and Swamy[127].

Ojha et al[128] carried out alkylation of phenol with tertbutyl alcohol (TBA) in batch mode over a zeolite catalyst, synthesized from fly ash by hydrothermal treatment. The effects of various parameters, such as reaction temperature, reactant ratio (molar ratio of phenol to tert-butyl alcohol) and catalyst loading oh the rate of reaction of phenol were studied with the synthesized catalyst. The alkylation reaction was found to be surface reaction controlled with negligible mass transfer resistance. An Langmuir-Hinshelwood-Hougen-Watson (L-H-H-W) surface reaction controlled kinetic model was developed and the model parameters were estimated. From the estimated kinetic constant at different temperatures, the activation energy of the phenol alkylation reaction was determined to be 37.3 kJ/mol.

A kinetic study of phenol alkylation with methanol in the presence of y-Al2O5 has been carried out. A mechanism the reaction of phenol and methanol adsorbed on acid base pair sites is proposed by Marczewski et al [129].

1.12 *Reason for undertaking the present problem*

Alkyl derivatives of hydrocarbons are important intermediates for many industries like petroleum, fine chemical and drug industries. Alkylaion of aniline produces N-ethylaniline and N,N-diethylaniline, former is an important dyestuff while latter finds application as co-catalyst in polymerization reaction, anticorrosive agent in acidic medium and as an antioxidant for lubricating oils. Alkylated phenols as o-cresol and 2,6-xylenol are industrially valuable compounds, which are utilized extensively as starting materials and intermediates in agrochemical and polymer industries. Pyrazine and its derivatives are industrially valuable compounds as intermediates for flavoring materials, drugs and agrochemicals. 2-methylpyrazine is the pivotal intermediate for obtaining 2-cnopyrazine, which when, hydrolyzed, yields pyrazinamide, which is a well known anti-tubercular drug. However natural availability of alkyl derivative is very limited therefore chemical routes are required for their synthesis.Zn-Co and Ni-Co ferrites have been shown to possess very catalytic activity for alkylation of aniline.However there seems to be no report on use of Mn-Cu ferrite as alkylating catalysts.In present work we have synthesized methyl derivative of aniline over Mn-Cu ferrite series,and perfected the process. There are reports on use of Palladized Zn-Cr oxide, ZSM-5, modified copper-chromite, binary catalyst based on oxides of Zn variable valance metals, Zn modified zeolites, Zn modified ferrite and $CuO/ZnO/SiO_2$ as catalysts for the synthesis of 2-MP from ED(Ethylene diamine) and PG(Propylene glycol). To the best of our knowledge, there is no report on kinetics of cyclization of ED and PG to produce 2-MP over alumina supported Mn ferrite. The present problem of kinetic study of synthesis of 2-MP from ED and PG was undertaken with a view to find suitable rate law and to predict mechanism of the reaction. Similarly there are few reports on kinetics of phenol alkylation over niobic acid, Mg-Al calcined hydrotalcites but no published work on kinetic study ofphenol methylation in vapour phase over Al-ferrite catalyst. Therefore the present work of kinetic study and mechanism evolution was undertaken with a view to find suitable rate law and to predict mechanism of methylation of phenol over this catalyst.

REFERENCES

1. Battacharya A. K. and Nandi D. K. Ind. Eng Chem Prod Pros Dev. **14** (1975) 162.
2. Narayanan S & Prasad B. P., J. Mol Catal. **96** (1995) 57.
3. Rice R. G. & Kohn E J., J. Am Chem Soc. **77** (1955) 4052.
4. Narayanan S. & Asima S. Stud Surf Sci Catal. **113** (1998) 667.
5. Elangovan S. P. Kannan C. Arabindoo B. & Murugeshan M. Appl. Catal A: General **174** (1998) 213.
6. Narayanan S. & Deshpande K., J. Mol Catal. **104** (1995) LI09.
7. Narayanan S. & Deshpande K. Appl. Catal: A. General **135** (1996) 125.
8. Narayanan S. & Deshpande K. Microporous Mater, **11** (1997) 77.
9. Narayanan S. & Dcshpande K. Stud Suif Sci Catal. **113** (1998) 773.
10. Narayanan S. & Dcshpandc K. Bull Catal. Soc. Ind. **9** (1990) 53.
11. Radhcshyam A. Dwivcdi R. Rcddy V. S. Chary K. V. R. & Prasad R., Green Chem. **4** (2002) 558.
12. Sreckumar K., Mathcws T., Rajgopal R., Vctrivcl R. & Rao B. S. Catal Lett. **65** (2000) 99.
13. Srcckumar K., Jyothi T. M., Mathcw T., Talawar M. B., Sugunan S. & Rao B. S., J Mol Catal. **159** (2000) 327.
14. Sreckumar K, Mathcw T., Dcvassy B. M., Rajgopal R., Vetrivel R. & Rao B. S., Appl Cata A: General **205** (2001) 11.
15. Forni L. and Pollesel P., J. Catal. **130** (1991) 403.
16. Kulkarni S. J., Subramanyam M., Rama Rao A. V., Indian J. Chem. **32A** (1993) 28.
17. Subramanyam M., Kulkarni S.J., Rama Rao A.V., Indian J. Chem., **2** (1995) 237.
18. Balpanov D. S., Krichevskii L. A., Kagarlitskii A. D., Russian Journal of applied chemistry **74,** 12 (2001) 2125.
19. Anand R., & Rao B. S., Catalysis communications **3** (2002) 29.
20. Anand R., Hegde S. G., Rao B. S., and Gopinath C. S., Catalysis communications **84** (2002) 265.

21. Ilnam Park, Jeongho Lee, Youngyoo Rhee, Yohan Han, Hyungrok Kim., Applied Catalysis, A: General **253** (2003) 249.
22. Balsama S., Beltrame P., Beltanne P. L., Fornil and Zuretti G., Appl. Catal **13** (1984) 161
23. Pierantozzi R. and Nordquist A. F., Appl. Catal **21** (1986) 263
24. Beltrame P., Beltanne P. L., Fomil L., Carniti P. and Castelli A., Appl. Catal **29** (1987) 327.
25. Santessaria E., Grasso D., Gelossa D. and Carra S., Appl Catal **64** (1990) 83-99.
26. Santessaria E., Diserio M., Ciambelli P., Gelosa D. and Carra S., Appl. Catal **64** (1990) 101.
27. Klemn L. H., Klofenstein C. E. and Shabati J., J. Org. Chem **35** (1970) 1069.
28. Nozaki F. and Kimura I., Bull Chem. Soc. Jpn **50** (1977) 614.
29. Tanabe K., Hattori H.., Sumiyoshi T., Tamaru, K. and Kondo T., J. Catal **53** (1978) 1.
30. Sato S., Koizumi K. and Nozaki F., Appl. Catal. **A133** (1995) L7.
31. Sato S., Koizumi K., and Nozaki F., J. Catal. **178** (1998) 264.
32. Narayanan S., VenkatRao V. and Durgakumari V., J. Mol. Catal **52** (1990) L29.
33. VenkatRao V., Durgakumari V. and Narayanan S., Appl. Catal. **49** (1989) 165.
34. VenkatRao V., Chary K. V. R., Durgakumari V. and Narayanan S., Appl Catal **61** (1990) 89.
35. Velu Sand Swamy C. S., Appl. Catal **A119** (1994) 241.
36. Velu Sand Swamy C. S., Appl. Catal. **A145** (1996) 141.
37. Velu Sand Swamy C. S., Appl. Catal. **A145** (1996) 225.
38. Iyothi T. M., Rao B. S., Sugunan Sand Sreekumar K., Ind. J. Chem **38A** (1999)
39. Sreekumar K. and Sugunan S., J. Mol. Catal. **A185** (2002) 259
40. Sreekumar K. and Sugunan S., Appl. Catal. **A230** (2002) 245
41. H. Kashiwagi Y., Fujiki and Enomoto S., Chem. Pharm. Bull. **30** (1982) 2575.
42. Kashiwagi H. and Enomoto S., Chem. Pharm. Bull. **30(2)** (1982) 404.
43. Chang C. D. and Perkins P. D., U. S. Pat. **4388461** (1983) 4388461.
44. McAtcer C. D., Brown C. D. and R. D. Davis, U. S. Pat. **5780635** (1998).
45. Chang C. D. and Lang W. H., U. S. Pat. **4220783** (1980).
46. Gleiter H.., Prog. Mater. Sci., **33** (1989) 223.
47. Siegel R. W., Ann. Rev. Mater. **21** (1991) 559.
48. Trudeau M. L., Ying J.Y., Nanostr. Mater. **7** (1996) 245.
49. Ying J. Y., Tschope A., Chem. Engg. J. **64** (1996) 225
50. Tschope A. S., Ying J. Y., Nanostr. Mater. **4** (1994) 617.
51. Tschope A. S., Liu W., Flytzani-Stephanopoulos M., Ying J.Y., J. Catal. **157** (1995) 42.
52. Tschope A. S., Ying J. Y., Chiang Y. M., Mater. Sci. Engg. **A204** (1995) 267.
53. Aksay I. A., Trau M., Manne S., Honmav, Yao N., Zhaov, Fenterv, Eisenberger M., Gruner S. M., Science **273** (1996) 892.
54. Vanleerlam G. C., Jacobs J. P., Brogersma H. H. Surf. Sci. **147** (1994) 294.
55. Vanleerlam G. C., Jacobs J. P., Brogersmav, 'Fundamental aspects of heterogeneous catalysis studies by particle beam' NATO-ASI series, plenum, New York **B265** (1991) 399
56. Auroux A., Gervasini A., J. Phys. Chem. **91** (1990) 6371.
57. Romeijn F. C., Philips Research Reports **18** (1953) 304.
58. Verwey E. J. W., Heilmann E. L., J. Chem. Phys. **15** (1947) 174.
59. Blasse G., Philips research Rep. Supplement **3** (1964) 40.
60. Smit J., Wijn H. P. J., Ferrite, Philips Technical Library, Eindhoven, The Netherlands (1959) 137.
61. Gorter E. W., Philips Res. Rept. **9** (1954) 302.
62. Verwey E. J. W., Boer F. De, Van Santen J. H., J. Chem. Phys. **16** (1948) 1091.
63. Vishwanathan B., Murthy V. R. K. (Eds.)., Ferrite Materials (Science and Technology).
64. Barth T. F. W., Posenjak E., Kristallographie Zs. **84** (1952) 325.
65. Datta R. K., Roy K., Nature **191** (1961) 169.
66. Hafner S., Laves F., Krist Z. **115** (1961) 321.
67. Bout I., Tailhades P., Rousset A., Kannan K. R., Verelst M., Kulkarni G. U., Rao C. N. R., J. Solid State. Chem. **102** (1993) 414.
68. Large M., Kacim S., J. Solid State Chem. **125** (1996) 7.
69. Kester B. G., Pariat P., Dufour P., J. Solid State Chem. **126** (1996) 7.
70. Smith J., Wijn H. P. J., 'Ferrites' John Wiley & sons, N.Y. (1959).
71. Tikadzumi S., 'Physics of Ferromagnetism', Tokyo (1978).
72. Guillard C.H., Creveaux I., Cr. Acad. Sci. (Paris) **229** (1949) 1458.
73. Morish A. H., Clark P. E., Phys. Rev., **B11** (1975) 278.

74. Somorjai G. A., Surface Chemistry And catalysis, John Wiley & Sons, Inc., New York, (1994).
75. Somorjai G. A., Pure & Appl. Chem. **50** (1978) 963.
76. Klabunde K. J., Li Y. X., Khaleel A., Nanophase Materials (Eds. G. C. Hadjipanayis, R. W. Siegel) Kluwer Academic Publishers, The Netherlands (1994) 757.
77. Omata K., Takada T., Kasahara S., Yamada M., Appl.Catal. **146** (1996) 255.
78. Ghose J., Murthy K. S. R. C., J. Catal. **162** (1996) 359.
79. Awe A. A., Miliades G., Vickerman J. C., J. Catal. **62** (1980) 202.
80. Murthy K. S. R. C., Ghose J., J. Catal. **147** (1994) 171.
81. Severino F., Brito J., Carias O., Lainc J., J. Catal. **102** (1986) 172.
82. Kehl W. L., Rennard R. J., Patent U. S. **3** (1969) 450, 787.
83. Kung H. H., Kung M. C., J. Phys. Chem. **84** (1980) 383.
84. Sloczynski J., Ziolkowski J., Grzybowska B., Grabowski R., Wcislo K., Gengembre L., J. Catal. **187** (1999) 410.
85. Finocchio E., Busca G., Lorenzelli V., Willey R. J., J. Catal. **151** (1995) 204.
86. Cares W. R., Hightower J. W., J. Catal. **23** (1971) 193.
87. Rennard R. J., Kehl W. L., J. Catal. **21** (1971) 282.
88. Shangguan W. F., Teraoka Y., Kagawa S., Appl. Catal. **B16** (1998) 149.
89. Shangguan W. F., Teraoka Y., Kagawa S., Appl. Catal. **B8** (1996) 217.
90. John Jebarathinam N., Eswaramoorthy M., Krishnasamy V., Appl. Catal. **145** (1996) 57.
91. Chen W. S., Lee M. D., Lee. J. F., Appl. Catal. **83** (1992) 201.
92. Sreekumar K., Ph.D Thesis, Submitted to Cochin University of Science and Technology, Kochi, February (1999)
93. Dube G. R., Darshane V. S., J. Mol. Catal. **79** (1993) 285.
94. Sreekumar K., Thomas M., Jyothi T. M., Biju M. D., Sugunan S., Rao B. S., Pol. J Chem. **74** (2000) 509.
95. Roesky R., Weiguny J., Bestgen H., Dingerdissen U., Appl. Catal. **A176** (1999) 213.
96. Xanthopoulou G., Appl. Catal. **A182** (1999) 285.
97. Sloczynski J., Jans J., Machej T., Rynkowski J., Stoch J., Appl. Catal. **B24** (2000) 45.
98. Fierro G., Morpurgo S., Jacono M. L., Inversi M., Pettiti I., Appl. Catal. **A166** (1998) 407.
99. Sato S., Iijima M., Nakayama T., Sodesawa T., Nozaki F., J. Catal. **169** (1997) 447.
100. Castiglioni G. L., Vaccari A., Fierro G., Inversi M., Jacono M. Lo, Minclli G., Pettiti I., Porta P., Gazzano M., Appl. Catal. **A123** (1995) 123.
101. Miki J., Asanuma M., Tachibana Y., Shikada T., J. Catal. **151** (1995) 323.
102. Ramankutty C. G., Sugunan S., Appl. Catal. **A 218** (2001) 39.
103. Sreekumar K., Mathew T., Mirajkar S. P., Sugunan S., Rao B. S., Appl. Catal **A 201** (2000) L1.
104. Sreekumar K., Jyothi T. M., Mathew T., Talawar M. B., Sugunan S., Rao B. S., J. Mol. Catal. **A159** (2000) 327.
105. Sreekumar K., Raja T., Kiran B. P., Sugunan S., Rao B. S., Appl. Catal. **A 182** (1999) 327.
106. Sreekumar K., Jyothi T. M., Talawar M. B., Kiran B. P., Rao B. S., Sugunan S., J. Mol. Catal. **A 152** (2000) 225.
107. Pierantozzi R., Nordquist A. F. Appl. Catal. **21** (1996) 263
108. Belitram P., Belitram P. L., Carniti P., Forni L., Appl. Catal. **29** (1987) 327
109. Samolada M. C., Grigoriadou, Kiparissides Z., Vasalos, I. A., J. Catal. **152** (1995) 52
110. Sreekumar K., Mathew T., Rajgopal R., Vetrivel R., Rao B. S. Catal. Lett. **65** (2000) 99
111. Sreekumar K., Mathew T., Devassy B. M., Rajgopal R., Vetrivel R., Rao B. S., Appl. Catal.A: **205** (2001) 11.
112. Sreekumar K., Sugunan S., Appl. Catal. A: **230** (2002) 245
113. Sreekumar K., Sugunan S., J. Mol. Catal. A: **185** (2002) 259
114. Reddy V. S., Radheshyam A., Dwivedi R., Prasad R., JCTB, **79** (2004) 1057.
115. Russell Drago, Physical Methods in Chemistry, Saunders College Publishing.
116. Brunauer S., Emmett P. H. and Teller E., J. Am. Chem. Soc. (1938) 309.
117. Dong Xue, Liu Jing, Shi Lei and Sun Qi, Journal of Natural Gas Chemistry **17** (2008) 110.
118. Bankupalli, Satyavathi, Kotra, Viswanath, Cheedipudi, Vijaya Lakshmi, and Venkateshwar S., International Journal of Chemical Reactor Engineering **6** (2008) A57.
119. Nishamol K., Rahna K. S., Sugunan S., Journal of Molecular Catalysis. A, Chemical **209** (2004) 89.
120. Frank B., Habel D., Schomacker R., Catalysis Letters **100** (2005)181.
121. Yuvraj S, and Palanichamy M., React Kinet Cseal Lete. **57** (1996) 159.
122. Narayanan S., Durga Kumari V. and Sudhakar Rao A. Applied Catalysis A: General **111**, 133
123. Kirichenko G. N., Glazunova V. I., Balaev A. V., and Dzhemilev Petroleum Chemistry **48** (2008) 389.
124. Charry K. V. R., Ramesh K., Vidyasagar G. and Venkat Rao V., J. Molecular Catalysis A. **198** (2003)195.
125. Sarla Devi G., Giridhar D. and Reddy B. M., Journal of molecular catalysis. A: Chemical **181** (2002) 173.
126. Bal Rajaram and Sivasanker Applied catalysis. A: General **246** (2003) 373.
127. Velu S. and Swamy C. S. Appllied Catalysis A: General **145** (1996) 225.

128. Ojha Keka, Pradhan Narayan C. and Samanta Amar nath Chemical Engineering Journal **112** (2005) 109.
129. Marczewski M., Perot G. and Guisnet React Kinet Catal Lett. **57** (1996) 21.

Table 1.1

Type	Structure	Examples
Normal	$(A^{+2})[B_2^{+3}]O_4$	$ZnFe_2O_4$, $ZnCrFe_2O_4$, $ZnCr_2O_4$, $MgCr_2O_4$
Inverse	$(B^{+3})[A^{+2}B^{+3}]O_4$	$MgFe_2O_4$, $NiFe_2O_4$, CO_2O_4
Random or mixed	$(A^{+2}_y B^{+3}_v)[A^{+2}_{1-x}B^{+3}_{2-v}]O_4$	$MgCrFeO_4$

Table 1.2: Target materials and their K_α lines

Element	K_α Wavelength Å
Mo	0.7107
Cu	1.5418
Co	1.7902
Fe	1.9373
Cr	2.2909

Table 1.3: Comparison of interplaner distances, d and corresponding intensities for standard $ZnFe_2O_4$ and $MnFe_2O_4$ ferrites with synthesized $ZnFe_2O_4$ and $MnFe_2O_4$

ASTM 22-1012		ASTM 10-319		Experimental $ZnFe_2O_4$ & $MnFe_2O_4$				hkl
$ZnFe_2O_4$		$MnFe_2O_4$		$ZnFe_2O_4$		$MnFe_2O_4$		
d (A°)	I/I_{max} %	d (A°)	I/I_{max} %	d (A°)	I/I_{max} %	d (A°)	I/I_{max} %	
4.873	7	4.906	20	4.889	10	4.889	14	111
2.984	35	3.005	35	2.979	35	2.989	36	220
2.543	100	2.563	100	2.540	100	2.552	100	311
2.109	17	2.124	25	2.108	18	2.114	22	400
1.723	12	1.734	20	1.718	13	1.726	25	422
1.624	30	1.636	35	1.623	30	1.628	21	511
1.491	35	1.503	40	1.490	38	1.495	31	440
1.287	9	1.296	20	1.286	10	1.290	7	533
1.099	11	1.106	30	1.093	10	1.093	10	553
-	-	0.982	-	-	-	-	-	-

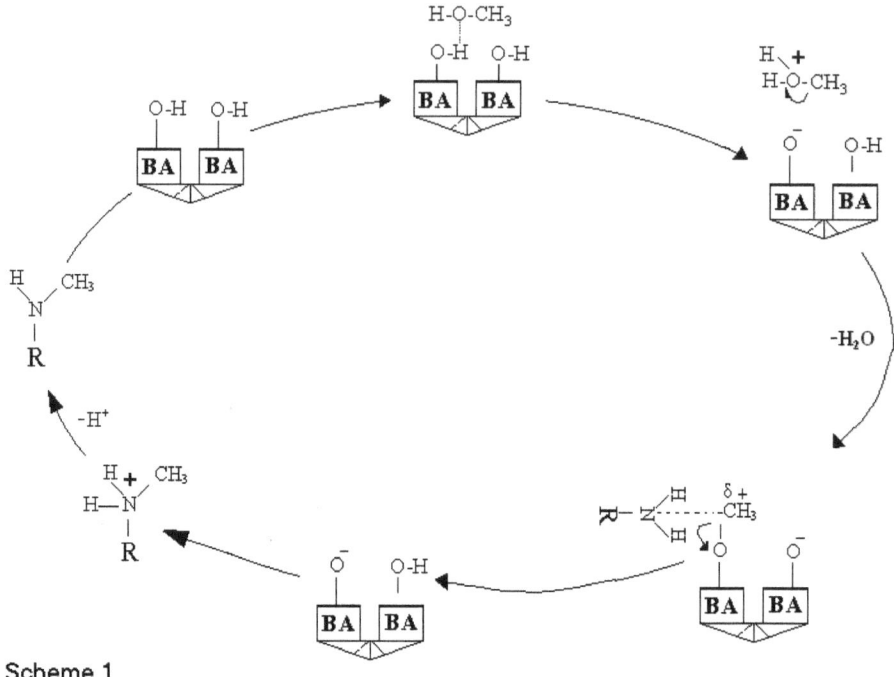

Scheme 1

Fig. 1.1: Mechanism of Aniline methylation over a catalyst possessing Bronsted sites

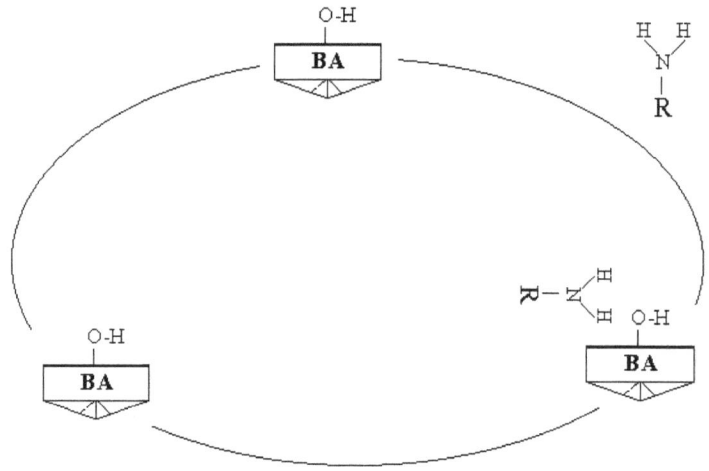

Scheme 1.1 (b)

Fig. 1.2: Formation of Anilinium ion of during the course of Alkylation

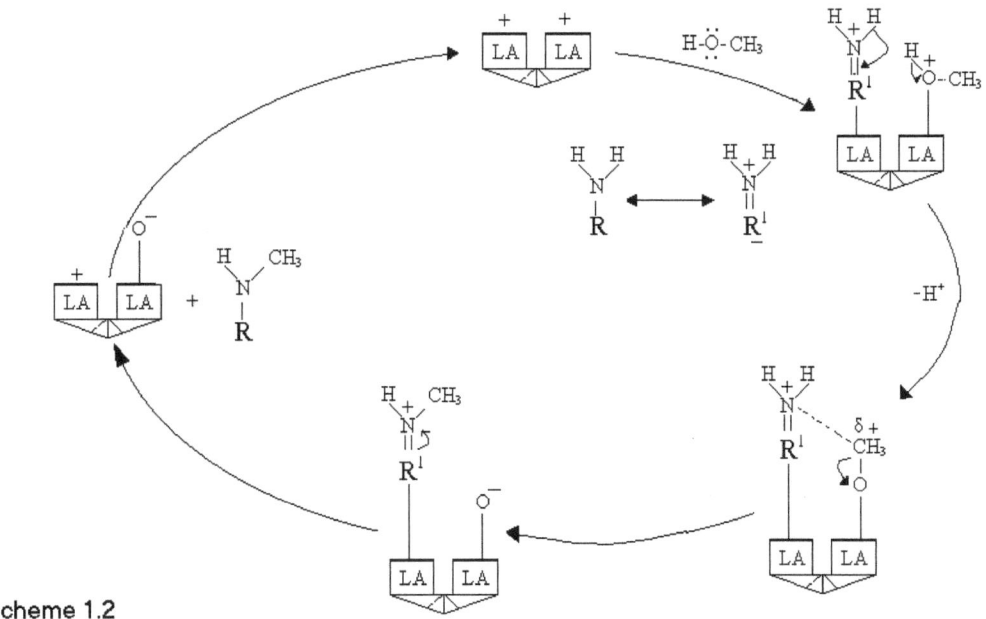

Scheme 1.2

Fig. 1.3 : Mechanism of Aniline methylation over a catalyst possessing Lewis Sites

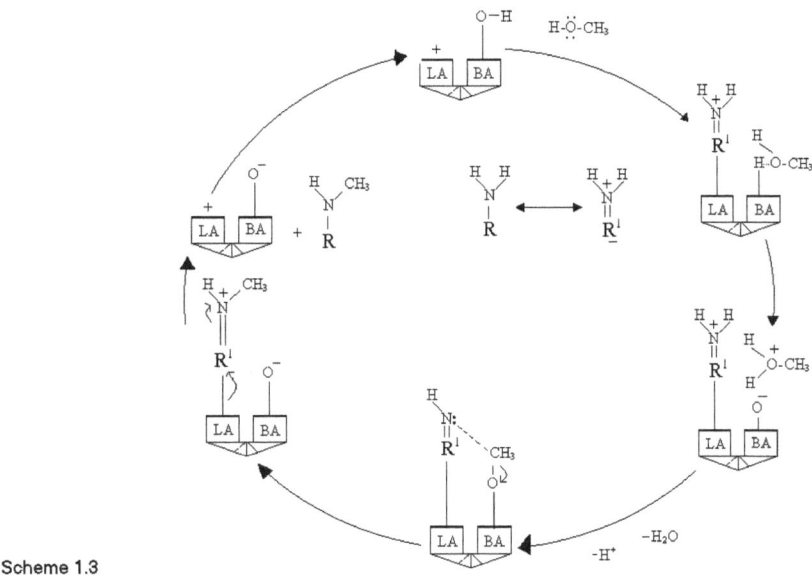

Scheme 1.3

Fig. 1.4 : Mechanism of Aniline methylation over a catalyst possessing both Lewis and Bronsted Sites

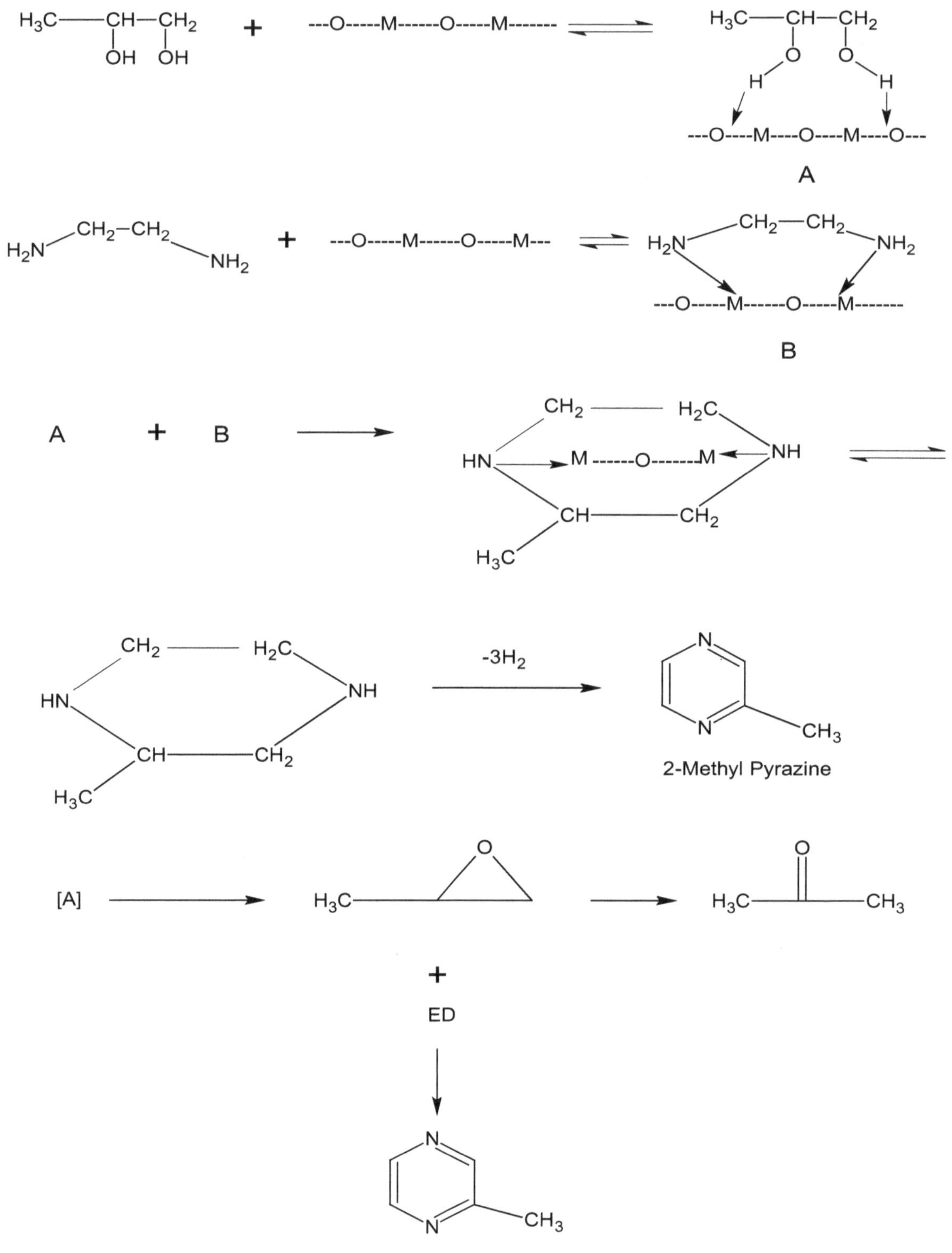

Fig. 1.5 : Mechanism of formation of 2-Methylpyrazine over a catalyst using Lewis Sites

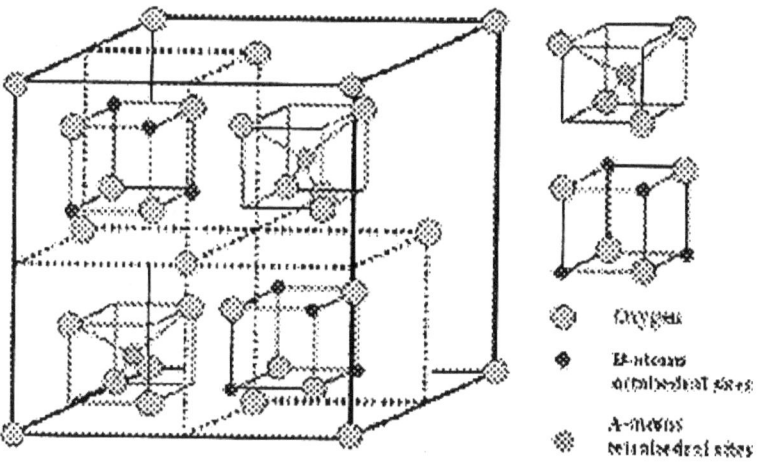

Figure 1.6 : Unit Cell of Spinel Structure

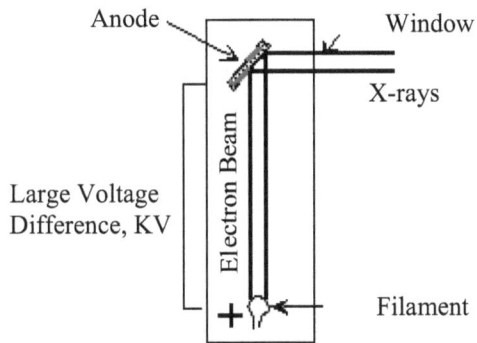

Fig.1.7 : Mechanism of X-ray production

$n\lambda = 2d\sin\theta$

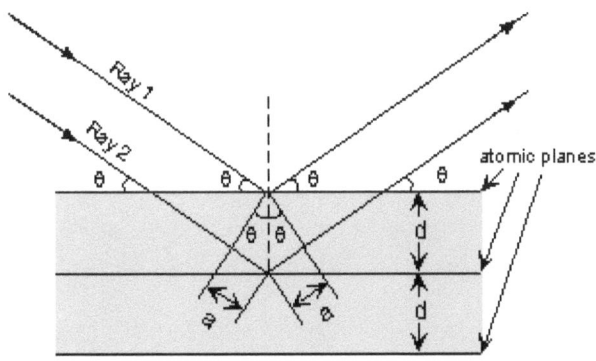

Fig. 1.8 : Scheme for Braggs equation

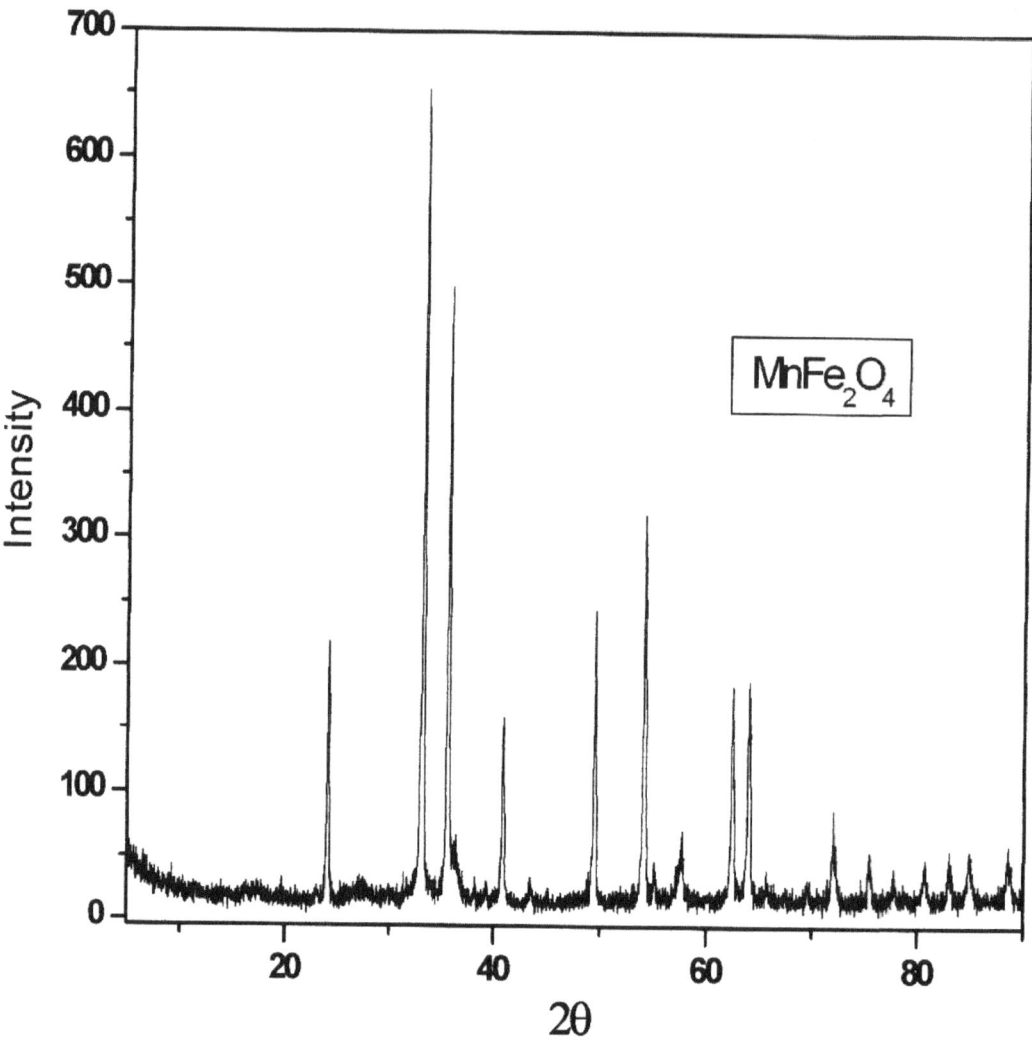

Fig. 1.9 : XRD pattern of MnFe$_2$O$_4$

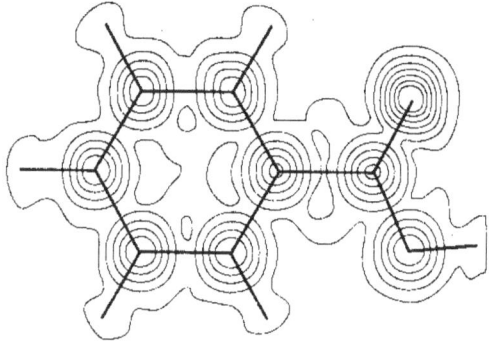

Fig. 1.10 : Electron Density map of benzoic acid

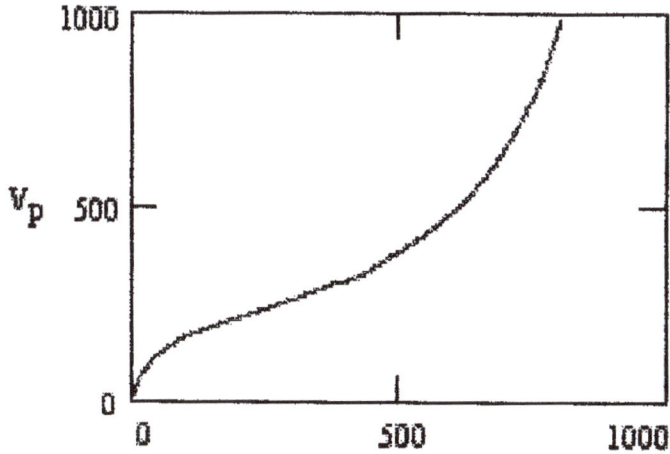

Fig. 1.11 : Nature of adsorption isotherm of gases near their condensation point

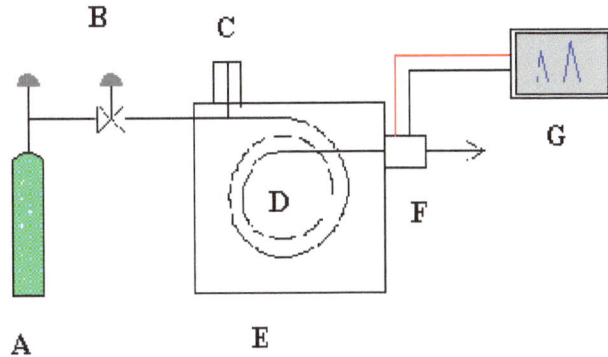

Fig. 1.12. : A schematic diagram of the GC apparatus. A = carrier gas cylinder, B = Flow controller, C= Injector port, D= Column, E Oven, F= Detector, G = recorder

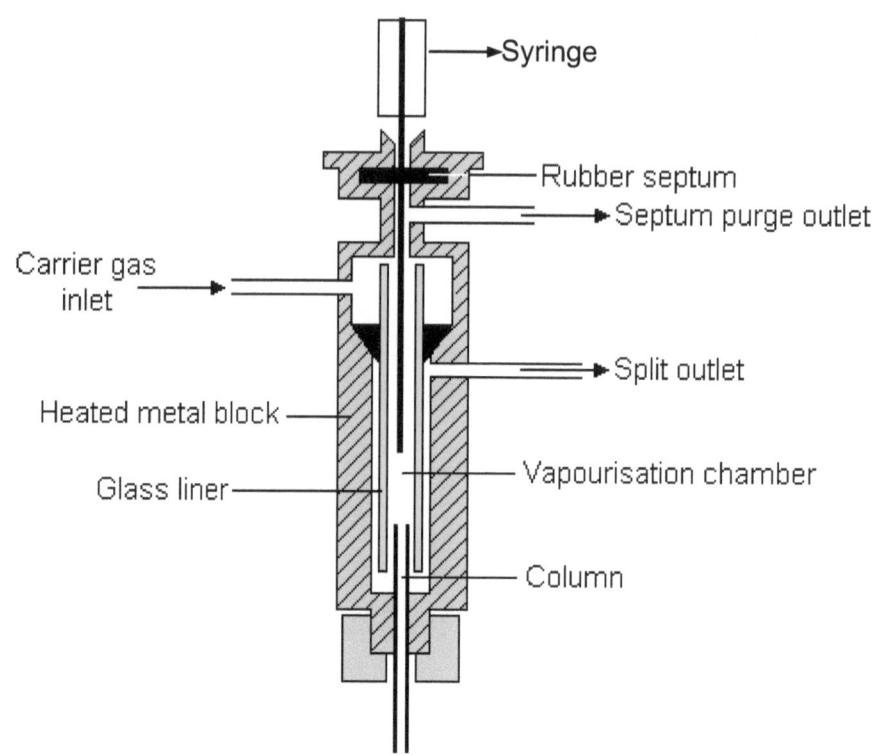

Fig. 1.13. : Schematic diagram of a split/splitless injector

Chapter - 2

2.1 The Catalytic Unit

The catalytic assembly is shown in Fig. 2.1. It consisted of 3 parts, the feed system, the reactor and the product recovery system. The feed system was made up of a calibrated tube from which the reaction mixture was allowed to drop under gravity at a known constant rate. The rate of flow was controlled with the help of a adjustable valve. At a time 1 to 2 g of reaction mixture was taken in the tube. The feed rate was in general $1 gh^{-1}$. This pump was connected to the reactor through a T tube. One side of the tube was connected to the feed pump, opposite side to the reactor though a standard glass joint and the third side of the T tube was connected to a nitrogen gas cylinder. Nitrogen gas functioned as a carrier gas to keep the down flow of reaction mixture.

The Reactor. It was fabricated from a pyrex glass tube about 1.5 'in length and 1'' in diameter. The upper part was fitted with a standard glass joint so that it can be attached to the feed system. The lower end was attached to a .8cm internal diameter tube, so that reactor could be attached to the product recovery system through a rubber tube connection. A concentric tube of diameter 0.8 cm, closed at one end, was joint to the bottom of the reactor which reached up to the bottom of the catalyst packing. This tube was used for inserting a chrom-alumel thermocouple to measure the temperature of the catalyst bed.

The reactor was heated by two tubular furnaces. The upper furnace along with the upper part of the reactor functioned as preheater and was used to vaporize the reactants. The lower furnace and lower part of reactor contained the reaction zone, where catalyst was housed.

The product recovery unit. This unit consitaed of a cold water condensed, and a product collector. The product collector was also kept in ice-salt bath. The product collector was open to atmoshphere. A long rubber tube was connected to this opening, to send he exhaust gases out-of- the laboratory.

2.2 Preparation of Ferrite Catalysts

All chemicals were of LR grade and were used without further purification. Distilled water was used for making solutions and as well as for washings. Calculated amounts of metal salts were dissolved in large excess of water and their hydroxides were co-precipitated under constant stirring by adding dilute solution of NaOH. The precipitate was further digested for 1 h at 80 °C. The precipitate was washed by repeated decantation till free from anions. The catalysts were sized by passing them through standard sieves. Finally, the precipitate was filtered, oven died and calcined at 500°C. The catalysts were sized by passing them through standard sieves.

2.3 Measurement of Acidity by Ammonia Desorption Method

Acidity measurement were made by ammonia desorption method. In this method acidity of the catalyst is determined using ammonia as adsorbate. 5g of the catalyst was packed in a Pyrex tube down flow reactor and heated to 673 K under nitrogen gas flow rate of $1 Lh^{-1}$ for 3 h. The reactor was then cooled to 298 K and adsorption conducted at this temperature by exposing the sample to ammonia for 2 h. physically adsorbed ammonia was removed by purging the sample with a nitrogen gas flow rate of $2 Lh^{-1}$ at 353 K for 1 h. The acid strength distribution was obtained by raising the catalyst temperature to 773 K in a flow of nitrogen gas of $2 Lh^{-1}$ and absorbing the ammonia evolved in 0.1 N HCl. Quantitative estimation was made by titrating the untreated acid with standard 0.1 M NaOH solution in different temperature ranges using phenolphthalein as indicator.

2.4 GLC Analysis of Reaction Mixture

The liquid samples were subjected to analysis using a Gas Liquid Chromatography (Chemito Gas Chromatograph 7610) equipped with Flame Ionization Detector and SE-30 column. Nitrogen was used as a carrier gas with flow rate of $3 \times 10^{-5} m^3$ / min. The equipment was programmed for analysis of samples in different temperature ranges. Parameters used for analysis of reaction mixture obtained from alkylation of aniline and cyclization of propylene glycol and ethylenediamine to 2-methylpyrazine are listed in Table 3. Cooling parameters were same for both the analysis. The nitrogen, hydrogen and air pressure were kept respectively at 1.5, 1.0 and 0.6 bar. Qualitative analysis was made by comparing the retention time of individual chemical with that of standard chemical. Quantitative analysis was made from the peak area of the particular chemical.

A software C- Chemitogram-2000 was used for quantitative analysis as well as qualitative analysis. A dedicated computer was used to operate the machine as well as for preparation of results. A photograph of the equipment is presented in picture 3.1.

Table : GLC programming parameters for different samples.

Program No.	Reaction	Oven Temperature range	Injector Temperature	Detector Temperature
		°C	°C	°C
1.	Aniline alkylation	130-200	200	200
2.	Cyclization	100-200	180	200
3.	Cooling	40	40	40

Yield, selectivity and rate of the reaction were computed from the GLC peak areas as follows.

Yield =

Selectivity =

Rate of reaction =

2.5 The XRD Recording

The XRD measurements were made by using Rigaku X-ray powder diffractometer equipped with graphite–crystal monochromator (for the diffracted beam) and scintillation counter. CuKα radiation with wavelength 1.5406 Å was used in 2θ ranges 0-90. The poder sample was mounted on a glass plate. The position of source was fixed while sample holder and recorder were made to rotate simultaneously with angle of θ and 2θ respectively. The phase purity of the samples were confirmed by comparing the diffractogrames with those of standard samples available in JCPDS cars. Mean crystallite size was determined by measuring the broadening of the peak (311) and applying the Debye equation [3].

SEM recordings: The morphology of the best catalyst $MnFe_2O_4$ was observed with JEOL JSM-5600 scanning electron microscope (SEM) equipped with an Energy Dispersive X-Ray detector (EDX). The samples were prepared by evaporating a drop of the sample on a carbon – coated copper grid. All images were obtained in the SEM mode with the emission gun operated at 20kV acceleration voltage.

2.6 Mössbauer Measurements

The Mössbauer measurements were carried out in the conventional transmission geometry in the constant acceleration mode. The Wiessel make velocity drive, LND make proportional counter and the ORTEC – MCS was used in the system. The source used was Co-57 in Rh matrix of ~ 6mCi strength. The spectrometer is calibrated with respect to the natural iron. The spectrum is fitted with doublet using NORMOS-SITE least square routine for the hyperfine parameters.

Atomic Force Microscopy: The surface morphology was studied using AFM instrument of Digital instrument Inc. USA using Nanoscope III and in contact mode.

2.7 ESCA Measurements

The ESCA (XPS) measurements were performed with a spectrometer from VSW, (U.K.) using an Al-Kα radiation (1486.6 eV). The residual pressure inside the analysis chamber was 1.2×10^{-9} Torr. The spectrometer was calibrated by using the photoemission lines of Ag (Ag $3d_{5/2}$ = 367.9eV, with reference to the fermi level). The full width at half maximum (FWHM) was 1.0eV under the recording conditions. Peaks were recorded with constant pass energy of 40eV.

High purity samples were fixed on the sample holder in a glove box filled with a continuous ultra high purity N_2 atmosphere and directly fixed on to the introduction chamber of the spectrometer. Charge referencing was done by C1s line at 284.6eV from the carbon contamination layer. Charging effects were minimised with a low energy electron flood gun in conjunction with a transmitting fine mesh proximity screen. The XPS signals were analysed by using Origin 6.1 software program.

Sample calculation

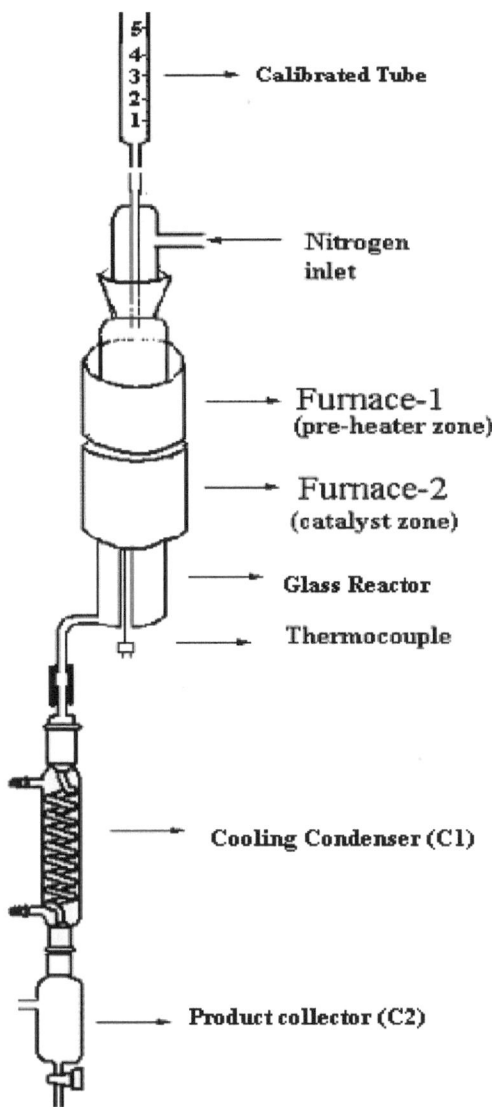

Experimental assembly for fixed bed reactor unit

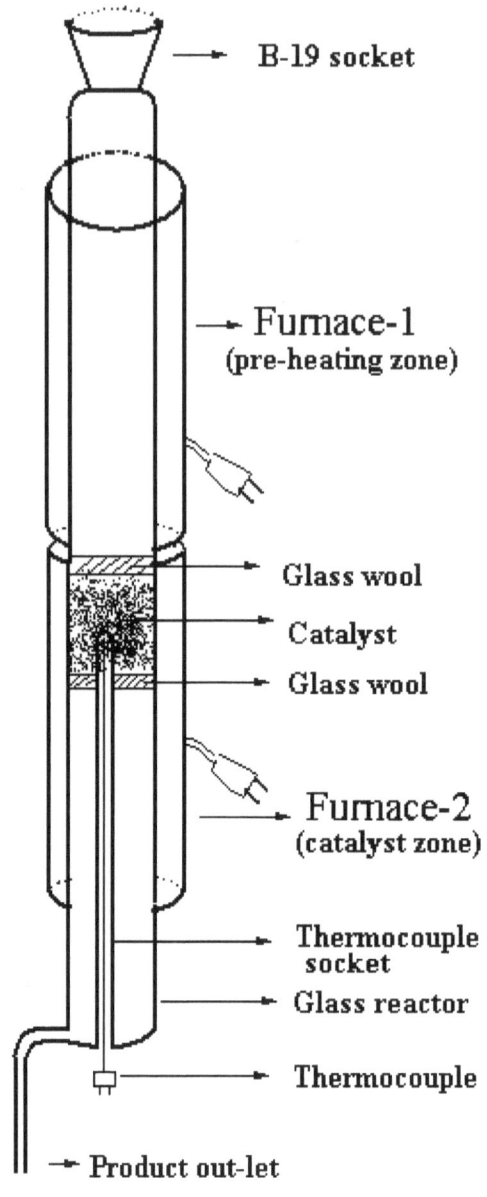

Schematic diagram of the reactor

XRD Measurements. The xrd measurements were made on a

Chapter - 3
Methylation of Aniline over Mn-Cu Ferrites Catalysts

I. Introduction

Alkyl anilines are valuable intermediates for the manufacture of pharmaceuticals, drugs, dyes and agrochemicals. These compounds may be prepared by catalytic alkylation of aniline with Various alkylating reagents: alcohols [1,2], dimethyl carbonate [3] and olefins [4]. Most of the projects carried out so far have used methanol as an alkylating reagent [5-9]. There are only a few reports available in the literature [10] mainly our own reference on aniline alkylation over zeolites using ethanol as an alkylating reagent We have also described the use of several oxides [11] mixed oxides [11, 12] clays [13, 14] and zeolites [15-17] for producing alkyl anilines selectively under different experimental conditions.

The unusual properties exhibited by nanoparticle and their promising technological applications have attracted much interest in recent years. [18]. This particles are shown to posses low saturation magnetization, Ms, enhance coercivity, H_c, and higher curie temperature, T_c, as compared to bulk materials, [19-24]. This phenomenon have been ascribed to random canting of surface spins [25-26], pinning of spins at particles surface or presence of a dead layer around magnetic core materials [27] and interparticle interactions [28]. It has been shown that certain ferrites with nanosized particles, posses metastable state and the cation distribution is different from that of bulk ferrites, where divalent metal occupies only tetrahedral, A, site and Fe^{+3} occupies only octahedral, B, site. In addition to these physical effects, nanoparticle ferrites are found to be excellent catalyst of alkylation and cyclization reactions with very high conversion and selectivity [29-30]. The high performances of ferrite cata1.yst have been attributed to cations distribution of particles size, lattice parameters and acido-basic characters of ferrite catalyst. To the best of our know1edge, there is no report on the methylation of aniline over copper-manganese ferrites (Cu-Mn ferrite). The present problem of the methylation of aniline over C-Mn ferrites was there fore undertaken with a view to optimize the Cu-Mn-Fe composition and process conditions in the catalyst for maximum conversion of aniline. Besides these catalysts have been characterized by XRD and Mossbauer spectroscopy" BET surface area and acidity measltrements. The cation distribution obtained from mossbauer study has been used to explain acidity and cata1ytic activity of the catalysts.

II. Experimental

2.1 Preparation of the sample

Nanoparticle of $Cu_{1-x}Mn_x Fe_2O_4$ (x = 0, 0.25, 0.5, 0.75, 1.0) were prepared by co-precipitation method. Required quantities of $MnCl_2$. Fe $(NO_3)_3.9H_2O$ and $Cu(NO)_3.3H_2O$ were dissolved in excess of distilled water. The pH of the solution was adjusted to 9 with dilute solution of NaOH.. Excess of water fast stirring and slow addition of NaOH solution is a must for getting small size particles. The resulting mixture was heated at a temperature of 333 K along with stirring for one hour. Then, it was kept for settling in the heating mode at 333 K fur half an hour for digestion. The product was washed by repeated decantation with 1.5 dm^3 portions of water until the supernatant is free of Cl^- (about 10-15 washings was required). The precipitate was filtered through Buchner funnel, dried in oven at 393 K and calcined at 573 K. The material was sieved through a 6/10 mesh sieve. After sieving the bigger size clusters were kept for acidity and catalytic activity evaluation while smaller size clusters obtained after sieving, were ground into fine powders and were used fur XRD and mossbauer experiment.

2.2 Characterization techniques

X-Ray diffraction pattern were recorded using Cu - K_α radiation (angle range 200 to 65°) as the source on a Rigaku X-Ray diffractometer. Diffraction patterns show the sharp lines corresponding to single-phase spinel for an the samples as the entire peaks match well with the characteristic reflections of the corresponding ferrite. The average particle size was measured by broadening of the (311) peak and applying Debye - Scherer equation. Lattice cell parameters were also calculated using Bragg's equation.

2.3 Mossbauer measurements

Samples were crushed to obtain fine powder in order to observe mossbauer spectra. The mossbauer absorber was prepared by spreading the paste of powdered sample and vacuum grease over perplex glass that is used as the sample holder. The absorption spectra were recorded in transmission geometry at room temperature with constant acceleration Mossbauer drive along with a 256 multichannel analyzer, using Austin Scince Inc, USA. Mossbauer. A single line source ^{57}Fe(Rh) with initial activity of 10 mCi was used. Several runs were taken in order to check the reproducibility of the spectra. Total counts collected were about 10^6 or more. The Mossbauer spectrometer was calibrated using a 0.001 inch enriched α-Fe foil. The outer lines are separated through 10.68 mm/s. This is an excellent agreement with an ideal absorption spectrum and calibration was done accordingly. One channel corresponds to 0.0628 mm/s. The experimental data were computer fitted using the least square-fitting program assuming Lorentzian line shape with χ^2 minimization technique. The solid line through the data points is the result of the computer fit of the data.

2.4 Acidity measurement

Acidity measurement were performed by ammonia desorption method. This experiment was carried out in order to measure the acidity of the ferrite catalyst using ammonia as an adsorbate.1 g of the catalyst was packed in between the plugs of glass wool in a Pyrex tube down flow bed reactor and heated to 773 K under a nitrogen gas flow rate of 0.5 cm^3/s for 1.5 hour. The reactor wag cooled to room temperature and adsorption was conducted at this temperature during which the sample was exposed to ammonia for 1.5 hours. In order to remove physically adsorbed ammonia from the purged sample again nitrogen gas was allowed to pass through it as the same rate at 353 K for 1 hour. The acid strength distribution was obtained raising the catalyst temperature from 353 K -773 K during the flow of nitrogen gas and absorbing the evolved ammonia in distilled water containing phenolphthalein as an indicator. Titrating the water solution with standard 0.1 M HCl solution in different temperature ranges also did quantitative estimation.

2.5 Catalytic activity

The alkylation unit consists of three parts i.e. feed system, reactor and product recovery unit. The reactor was charged with a known weight (5g) of catalyst and was clamped in the assembly holder. The tubular heater was also clamped in the stand. The reactor was then joined to the product recovery unit and syringe pump filled with known volume of reactant mixture. The catalyst was activated at 773 K by passing air and then brought down to desired temperature by cooling down in the current of N_2. The mixture of reactant was fed by a 10 ml syringe pump. The liquid product. was condensed with the help of cold-water condenser, a cold trap and was analyzed by shimadzu gas chromatography using SE-30 column and FID detector.

III. Results and Discussion

3.1 XRD analysis

XRD patterns shown in fig 1. reveals the formation of spinel phase of ferrite as all the composition showed the characteristic reflection of the spine1 phase. All the peaks of $MnFe_2O_4$ and $CuFe_2O_4$ matched well with those as reported for the same in JCPDS Card No.10-319; 3-864. The reflections at the planes (111), (220), (311), (222), (400), (422), (511), (440) can be detected with d values of 4.89, 2.97, 2.53, 2.42, 2.08, 1.71, 1.61, 1.49 A respectively. Accordingly the lattice cell parameters were also calculated that matched well with the values as reported in the JCPDS card no.s given above. Lattice cell parameter shows a linear decrease from 8.49 A. to 8.39 A with the introduction of Cu in $MnFe_2O_4$ as could be seen clearly from the fig 2 and may be attributed to the replacement of Mn^{2+} (0.80 A) by smaller Cu^{+2} (0.69 A). This also confirms the formation of solid solution. The mean crystallite size was calculated by broadening of the (311) diffraction peak and applying Scherer formula and the range is in between 7.2 nm - 20 nm. Surface area was also estimated from the crystallite size data assuming the crystallites to be of spherical shapes. They depend on the size of the particle as is clear from the reported values in Table 1. The average particle size and lattice cell parameter for the above mixed ferrite nanoparticles are reported in table 1.

3.2 Mossbauer

Mossbauer absorption spectra for the samp1es i.e. x = 0, 0.5, 1.0 were recorded at room temperature as shown in Fig.3. The spectra were fitted using NORMOS fitting program and hyperfine parameters e.g. isomer shift δ, quadrupole splitting, Δ, line width, Γ, and hyperfine field, H_{int}, thus obtained are listed in Table 2. The Mossbauer spectra exhibit broad magnetic hyperfine sp1it sextets. In case of $MnFe_2O_4$ an additional doublet is seen which can be attributed to super paramagnetic behavior exhibited by smaller size particles. The doublet was fitted with two sub spectra .The isomer shift was found to be 0.24-0.27 mm/s which can be attributed to Fe^{3+} ions in tetrahedral and octahedral sites. The sextets were fitted with two sub spectra. One sextet is due to Fe^{3+} ion at the tetrahedral A site and the outer sextet is due to Fe^{3+} at the octahedra1 B sites, which indicates the ferrimagnetic behavior of the samp1es. All of them shows isomer shift values in the range 0.22 mm/s to 0.34 mm/s, which could be attributed to the Fe^{3+} ions present at both the A and B sites of the spinel structure. This indicates that Mn^{2+} is also distributed between tetrahedral site and octahedral site. Similar metastable structure has been proposed by C Rath et al in nanocrystalline Mn ferrite. However, the case is different in bulk ferrites where Mn^{2+} prefers tetrahedral site while Fe^{3+} prefers octahedral site.

However, distribution of Mn^{2+} on both the sites has also been reported for $MnFe_2O_4$ nanoparticles [mahmoud]. On addition of Cu^{2+} which can occupy both the sites but prefers the octahedral site, slight broadening of the fine width and reduction in the intensity of the central lines is observed. These results agree with those reported earlier in the literature [PCCPI. The tetrahedral and octahedral positions of iron in the ferrospinels could also be distinguished on the basis of Magnetic hyperfine field, BHF. It has been mentioned that the internal BHF field experienced at A sites is smaller than that exerted at B sites: Usually, contribution of Fe^{2+} to the hyperfine field is significantly smaller than that of Fe^{3+} i.e. ferric ions. Also the isomer shift values for ions located at the tetrahedral site are generally smaller than those at octahedral site. It has also been reported that Fe^{3+} ions are at A site while Fe^{3+}/Fe^{2+} both can reside at B site alternately. The observed isomer shift value confirms the presence of Fe^{3+} ions at both the sites while no Fe^{2+} is present in any of the composition. The intensity ratio A and B sites gives the ratio of the no. of Fe atoms at A and B sites respectively. Quadrupole splitting has value near by zero for both the sites. As in the presence of strong magnetic interaction, the distf1.:bntion of qnadrapole interaction that arises from chemical disorder, produces an appreciable broadening of the individual Zeeman lines for both tetrahedral and octahedral patterns, but doesn't produces an observable quadrapole line shifts. Data extracted from the fits are reported in table 2.

3.3 Effect of acidity and process variable

The performance of various Mn-Cn ferrite catalysts along with their acidity, XRD surface area, particle size, and lattice constants is presented in Table 1. It can be seen that conversion of aniline as well as acidity of different catalysts increase with increase in copper content of the ferrites. Similar trend can be seen for surface area. The increase XRD surface is due to decrease in particle size of the sample with increase in Cu content. Consistency in lattice parameters of all the ferrites confirms them to possess same lattice type as concluded in section.

The order of catalytic activity of ferrospinels towards overall conversion was found to be CF> MCF-3>MCF-2>MCF-I>MF. It can be concluded from these resu1ts that. catalytic performance of the ferrites under consideration is proportional to surface area as well as acidity. An examination of the acidity reveals CF to be the better catalyst as its acidity value is larger and it also increases with increasing x as is clear from the reported values in table 3. It also shows CF to be the better catalyst, as its aniline conversion (%) is larger .

The effect of methanol! aniline mo1ar ratio on the performance of copper ferrite is depicted in Figure .The conversion increases at lower molar ratio and tends towards limiting conversion at higher molar ratio, Maximum conversion of 82 % of aniline was obtained at methanol/ aniline mo1ar ratio of 5 with NMA selectivity of 80 %.

The temperature effect. on alkylation for MF at constant molar ratio was also studied in the temp range 473-773 K and resu1ts are presented in table 5. Negligible conversion can be seen below 473 K whi1e the conversion effectively occurs in the range 673 K - 700 K. The best performance by the catalyst was shown at temperature 673 K with conversion of 62 and 15 % for N-methyl aniline and N-N-dimethyl aniline respectively, and selectivity of 79 and 62 % for NMA and NNDMA respectively. Conversions decreased due to charring and deposition of carbon on the catalyst surface at temperature higher than 673 K.]

IV. Conclusion

1. $Mn_{1-x}Cu_xFe_xO_4$ (x = 0., 0.25 0.5 0.75 1.0) mixed ferrite nanoparticles were obtained in a broad range of Cu concentration 0.0 < x <1.0 by coprecipetation and digestion method. Room temperature mossbauer spectra of these fine particles exhibit slightly high va1ue of hyperfine field and broadening of the zeeman spectral lines shows strong ferromagnetic behavior of the three compositions. The occupancy ratio between Fe cation at A and B sites of the spinel structured is deduced from the fitted data of the mossbauer spectra.
2. These mixed ferrite spinel systems were also studied for the alkylation of aniline using methanol as the alkylating agent. These systems effectively alkyl ate aniline to N-methyl aniline and N-N-dimethyl aniline under optimized reaction condition. Highest. activity is obtained for CF whereas MF and MCF-1 were mildly activated. Substitution of Cu in MF leads to the increase in va1ue of acidity and the aniline conversion. This behavior suggests for Mn-Cu mixed ferrite system to be a good catalyst.

References

1. Chang, J. R., Sheu, F. C., Cheng. Y.M., and Wu. J-c.,: Appl. Catal. (1987), 33, 39.
2. Yashima, T., Sakaguchi, Y., and Namba, S., : Proceedings of Seventh International Congress on Catalysis, Kodensha, Tokyo: Elsevier, Amsterdam, (1980), PA 52-1.
3. Borade, R. B., Halgeri, A. B., and Prasad Rao, T. S. R., : Proceedings of Seventh International Congress on Catalysis, Kondensha, Tokyo, Elsevier, Amsterdam (1986), 851 .

4. Young, Lb., Butler, S. A., and Kaeding, W. W.: J.Catal. 76 (1382), 418.
5. Namba, S., Lanka, A., and Yashimam, T., : Zolites, 3 (1983), 106.
6. Chen, N. Y. : J. Catal., 114(1988) 17.
7. Kmnar, R., and Ratnaswamy, P., : J. Catal., 116 (1989), 440.
8. Yaug, L., AizBen, V., and Quinhva, Y., Apll. Catal. 67 (1991).
9. Bautista, F. M., Blanco, A., Camlpleo, .J. M., Garcio, A.. Luna. D., Marinas, J. M., and Romario, A. A., Catalysis Letters, 26 (1994), 159-167.
10. Long, G. N., Pellet, R. J., and Rabo, J. A., : US Pat., $, 528, 414 (1985).
11. Rabo, J. A., Pet1et, R. J., Cough, P. G., and Shamshoum, E.G., in 'Zeolites as Catalysts, Sorbents and Detergents Biulders' Edited by H. G. Karge, and J. Weitkemp, Elsvier, Amsterdam (1989).
12. Das, J., Lohokare, S. P., Chakrabarthy, D.K.,: Ind. J. Chem., Vol. 1A, (1982), 742-746.
13. Benedtte. C., Emanuets, S., Marcello, G., and Fullvoi, T., : br. Pat. 8008323 (1981).
14. Aurel, L., gheorghe, I., Elena, m., and Vasilica, T. RO Pat. 71356.
15. Mc Allester, S. H., Anderson, J., and Buttlard, E. F., : Chem. Eng. prog. 43 (1947), 189.
16. Kaeding, W. W., and Hoand, R. E. : J. Catal., 190 (1988), 212.
17. Balakrishnan, I., Chunbhale, V. R., and Rao, B. S.,: 8th natl. symp. catal Vol. I. S. R. Naidu, B. K. Banerjee (Es.). PDPL, Sindri, (1987), 155.
18. Hadjjpanayis, G. C., Siegel, R. W., Nanophase Materials (Kluver, Dordreeht, 1994).
19. Gajbhiye, N. S., Met, Mater. Process, 10, (1998), 247.
20. Kodama, R. H., Berkowitz, A. E., mcniff, E. J., and Foner, S., Phys. Rev. Lett. 77, (1996), 394.
21. Chen, p., Sorensen, C. M., Klabund, K.J., Hadjipanayis, G. C., Delvin, E. and Kostikas, A., Phys. Rev. B-54, (1996), 9288.
22. Pal, M., Brahma, P., Chakravorty, D., Bhattacharya, D., and Maiti, H. S., J. Magn. Magn Mater.
23. B. D. Cullity, Elements of X-Ray Diffarction.
24. J. A. Toledo, P. Bosch, M. A., Valenzuela, A. Montoya, N.Nava, J. Mol. Cata. A.: Chemical 125 (1997) 53-62.
25. H. P. Klug. L. E. Alexander. X-Ray Diffraction procedures wiley, New York, NY, 1974, p no.562.
26. Ferrites, B. Viswanathan Karoly Lazar, Thomas Mathew, Zsuzsanna Koppany, Janos Megyeri, Violet Samul, Subhash. P. Mirajkar, Ballopragad S. Rao and Laszlo Guczi, Phys. Cem. 2002 4 (3530-3536).
27. Z. X. Tang, C. M. Sorensen, K. J. Klabunde, G. C. Hadjipanayis,, J. colloid and interface science, Vol. 146, No.l, Oct. 1991.
28. M. H. Mahmoud, H. H. Hamdeh, J. C. Ho, M. J. O. shea, J. C. Walker. J. Mag. Mag. Mat. 220 (2000) 139-146.

Table 1 : Particle size, Lattice parameter, acidity and catalytic activity of $Mn_{1-x}Cu_xFe_2O_4$ catalyst. Temperature = 673 K, Mole ratio = 5.0

Composition	Aniline conversion	Particle Size (nm)	XRD Surface area (m^2/g)	Lattice parameter (A^0)	Acidity (m mole/g)
MF	28	13.9	86.16	8.49	1.2
MCF-1	32.2	20.9	56.15	8.45	1.2
MCF-2	40.4	14.6	78.67	8.43	1.3
MCF-3	58.0	13.9	81.29	8.40	1.4
CF	80.5	7.2	155	8.39	1.5

Table 2 : Mossbauer parameters extracted from the fitting for the composition $Mn_{1-x}Cu_xFe_2O_4$. Where I.S-Isomer shift, Q.S- Quadrupole splitting, BHF- Magnetic hyperfine field, H.L. W-half line width, RI -Relative intensity for A and B sites

Composition		Site	I.S (±0.03 mm/s)	Q.S (±0.02 mm/s)	BHF (± I T)	H.L.W. (± 0.01)	RI (%)
$MnFe_2O_4$	[D]	A	0.27	1.07	-	0.48	12.50
	[D]	B	0.24	0.82	-	0.36	12.50
	[S]	A	0.38	-0.212	50.12	0.47	37.50
	[S]	B	0.14	-0.12	50.00	0.41	37.50
$Mn_{0.5}Cu_{0.5}Fe_2O_4$	[D]	A	0.25	0.015	47.77	0.68	38
	[S]	B	0.32	-0.22	51.90	0.45	62
$Cu_{0.5}Fe_2O_4$	[S]	A	0.22	-0.04	48.15	0.71	45
	[S]	B	0.29	-0.1	51.14	0.67	55

Table 3 : Performance of various catalysts in the alkylation of aniline. Methanol/Aniline molar ratio = 5; Temperature = 673 K, WHSV = 0.2 h^{-1}

Catalyst (%)	Aniline Conversion (%)	Product distribution		
		Aniline	NMA	NNDMA
MF	28.0	70.0	21.8	8.2
MCF-1	32.2	65.2	27.6	7.8
MCF-2	40.4	62.4	34.0	3.6
MCF-3	58.0	50.6	47.4	2.0
CF	80.5	19.4	79.4	1.2

Table 4 : Effect of mole ratio on alkylation of Aniline Temperature = 673 K, WHSV = 0.2 h^{-1}
Catalyst = CnFe$_2$O$_4$

Mole ratio	Aniline Conversion (%)	NMA (yield)	NNDMA (yield)	NMA (Selectivity)	NNDMA (Selectivity)
2.5	64.2	20.6	10.0	67.3	32.7
5.0	82	61.2	15.3	80.0	20.0
7.5	79.2	63.6	8.2	88.5	11.5
1.0	73.4	71.2	5.2	93.2	6.8

Table 5 : Effect of temperature on alkylation of Aniline
Methanol/aniline molar ratio = 5
WHSV = 0.2h^{-1}
Catalyst = CuFe$_2$O$_4$

Temperature	Aniline Conversion (%)	Product distribution (%)			
		Aniline	NMA	NNDMA	Others
473	06.0	96.0	-	-	4.0
573	38.2	51.2	20.8	22.0	6.0
673	80.1	23.2	62.4	14.4	-
773	42.8	39.8	42.2	18.0	-

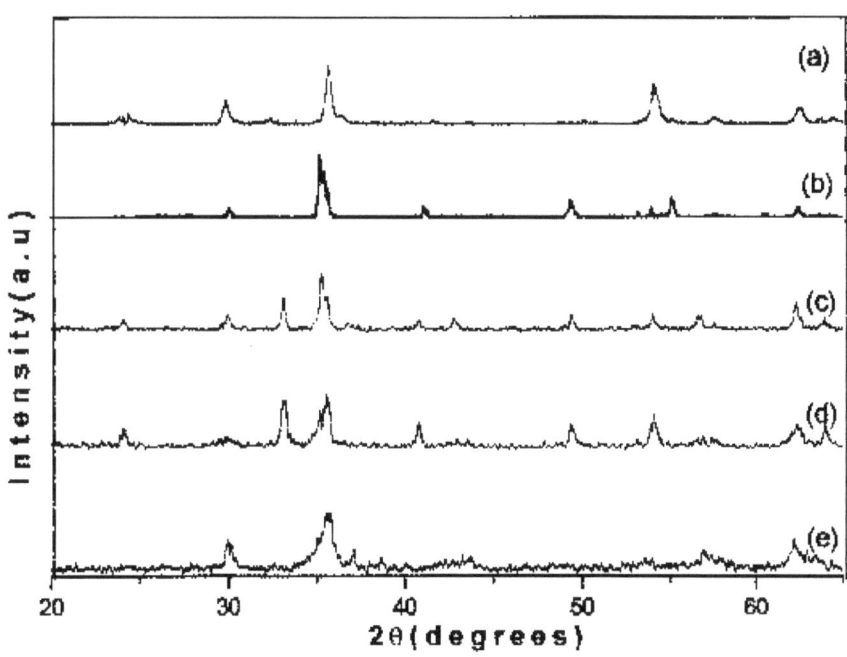

Fig. 1

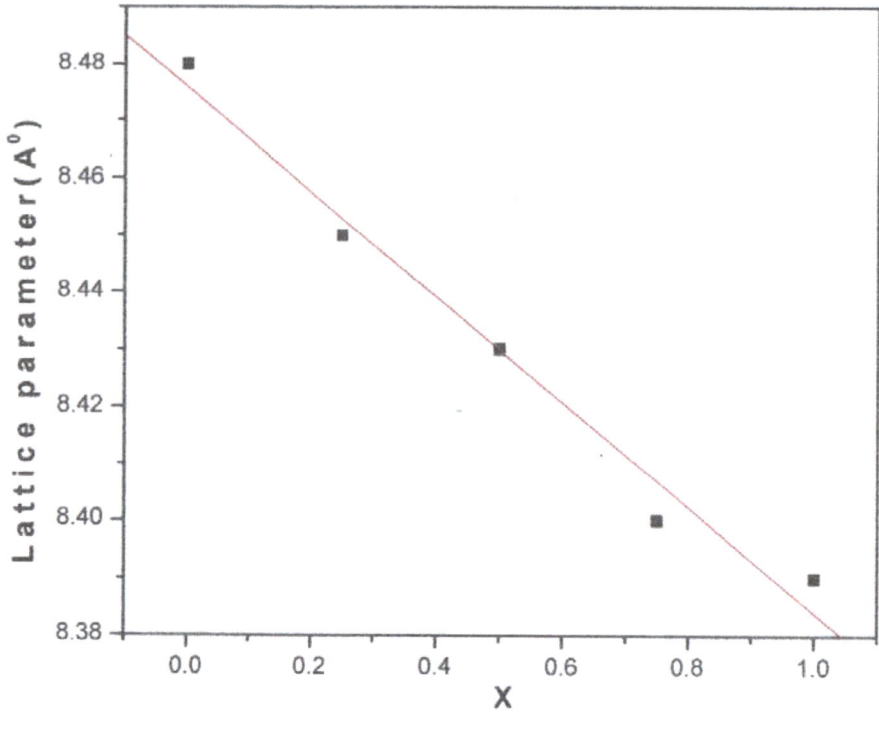

Fig. 2

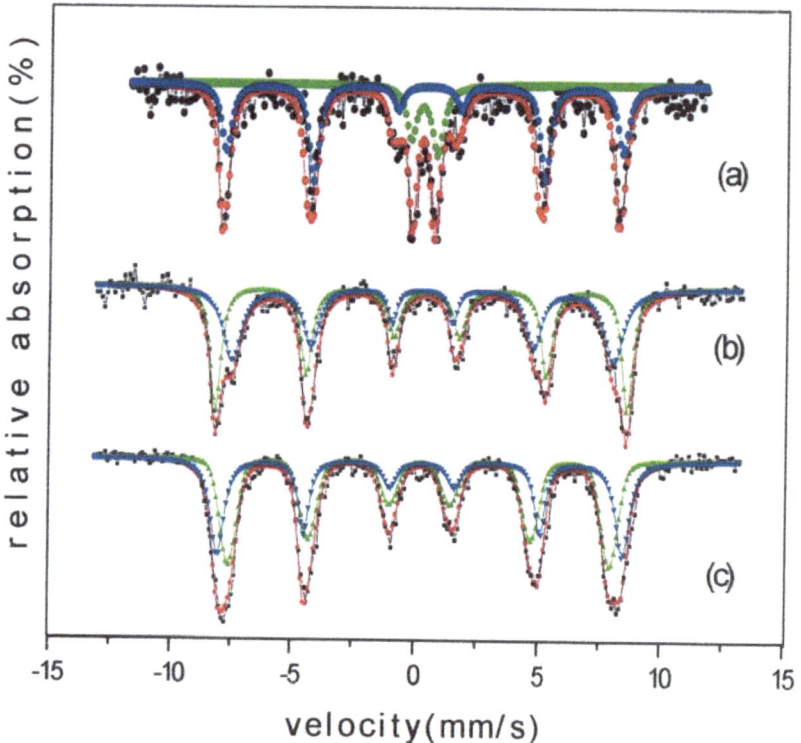

Fig. 3

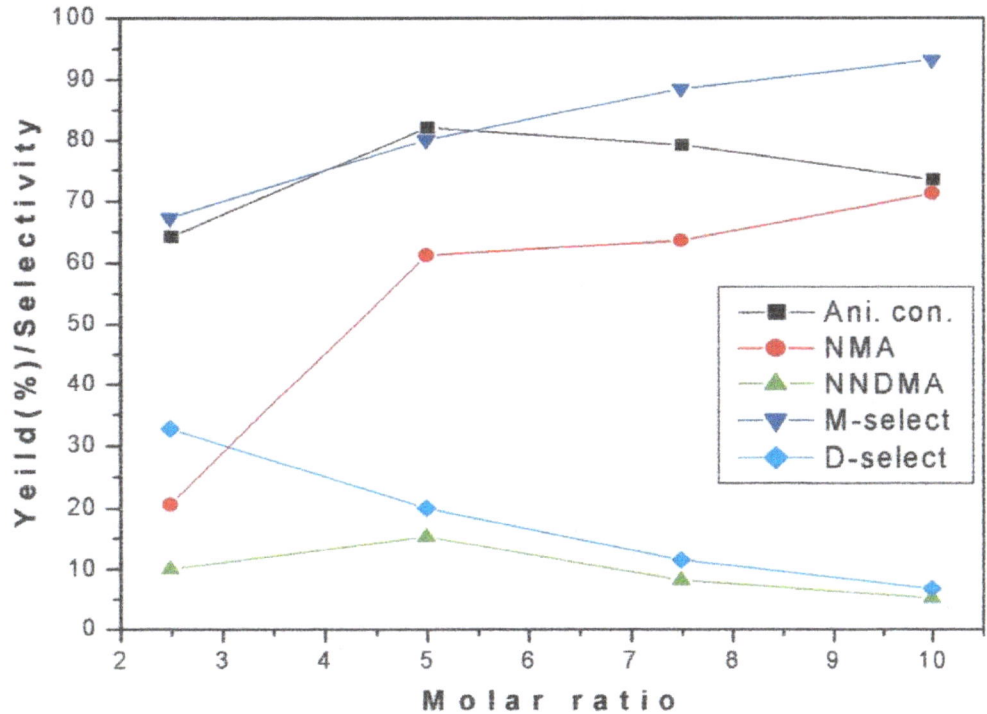

Fig. 4

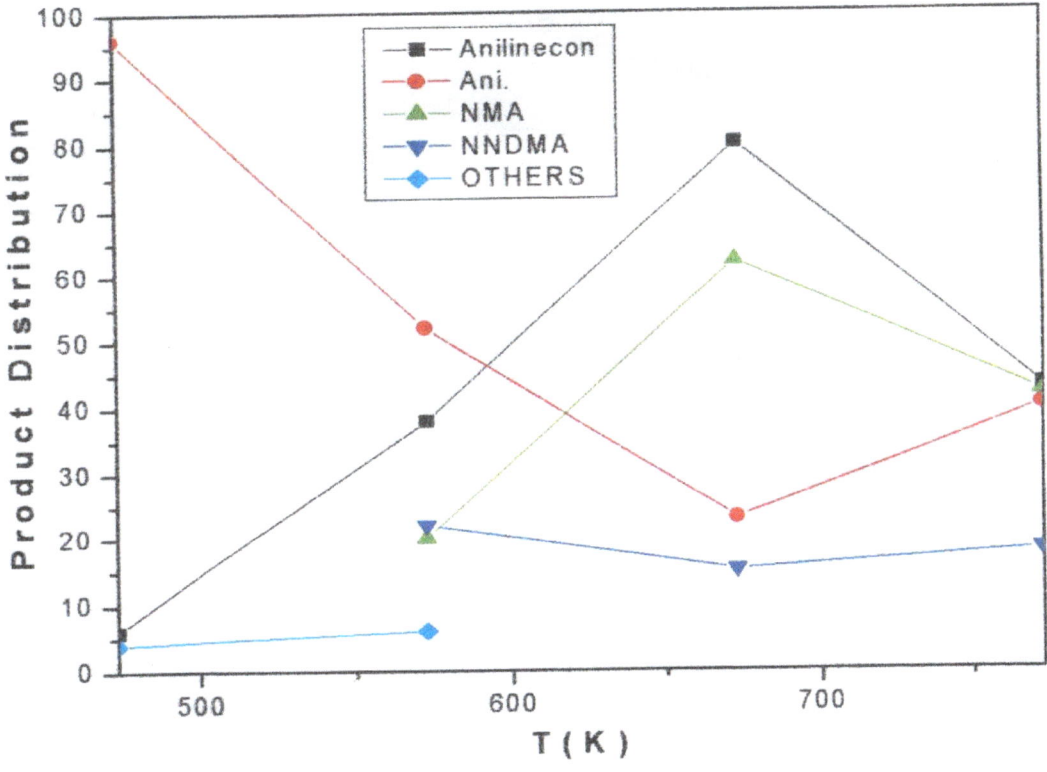

Fig. 5

Chapter - 4
Kinetics Studies and Mechanism Evolution of the Cyclization of Ethylene Diamine and Propylene Glycol over Alumina Supported Mn-Ferrite

Abstract- Kinetics of Cyclization of propylene glycol and ethylene diamine to 2-methyl pyrazine has been studied over alumina-supported nanocrystalline $MnFe_2O_4$ catalyst in a differential flow fixed bed reactor in the temperature range 200 – 300 °C. The partial pressures of propylene glycol and ethylene diamine were varied and rates were measured for the formation of 2 Methyl Pyrazine. Product selectivity as well as rate of formation of 2-methyl pyrazine were influenced by partial pressure of reactants. The rate equation $R = k\, K_G K_E (P_P)(P_E) / (1 + K_P P_P + K_E P_E)^2$ deduced, on the basis of adsorption of PG and ED in gas phase represented the data most satisfactorily.

Notation

R = Rate of formation of 2-methylpyrazine (μ moles / m^2 /h)

k = Rate constant,

K_G, K_E = Adsorption equilibrium constant,

P_P = Partial pressure of propylene glycol,

P_E = Partial pressure of ethylenediamine

I. Introduction

Nanosize materials are known to exhibit certain properties that are different from their bulk counterparts, (Interrante et al., 1998). Such materials possess higher surface area, band gap, coercivity, acedity, alkalinity and coordination of atoms. The chemical properties also vary because of the changes in the electron density as a function of particle size, resulting thereby in binding modes of the adsorbate molecules that are different than that observed in the case of corresponding bulk materials. Additionally, in case of supported catalysts, the electronic interaction between the nano-dispersed metal crystallites and the support materials are known to influence the chemisorptions properties. Providing support to the nanocrystalline metal oxides avoids their agglomeration in actual reaction conditions and helps achieve highly dispersed and uniform size particles (Anpo et al., 2003; Maira et al., 2000). In addition to large surfaces, support helps in the shape selectivity and also provides an inert envelope to protect from the chemical effect of the reaction medium.

2-MP is the pivotal intermediate for obtaining 2-cyanopyrazine, which on hydrolysis yields pyrazinamide, a well-known anti-tubercular drug. Conventionally, chromite catalysts are used for vapour phase synthesis of 2-MP. There are reports on the use of Palladized Zn-Cr oxide (Forni et al., 1991), ZSM-5 (Kulkarni et al., 1993), modified copper-chromite (Ilnam et al., 2003), binary catalyst based on oxides of Zn and variable valence metals (Balpanov et al., 2001), Zn-modified zeolites (Anand et al., 2002), Zn-modified ferrite (Anand et al., 2002) and $CuO/ZnO/SiO_2$ (Subramanyam et al., 1995), as catalysts for synthesis of 2-MP from ED and PG. To the best of our knowledge, there is no report on the kinetics of cyclization of propylene glycol (PG) and ethylene diamine (ED) to produce 2-MP over alumina supported nanocrystalline manganese ferrite. The present problem of kinetic study of synthesis of 2-MP from PG and ED was therefore undertaken with a view (1) collect data on the kinetics of the vapour phase synthesis of 2MP over alumina support Mn-ferrite (AMF) catalyst (2) to find a suitable rate law, which can explain the data satisfactorily, and (3) to predict a mechanism of the reaction. Besides, catalyst has been characterized using IR, XRD, Mössbauer, SEM, Raman, Surface Area and Ammonia desorption methods with the objective of understanding structure and nature of bonding over catalyst surface.

In the present investigation of the kinetics of 2MP synthesis over alumina supported catalyst, we have followed the usual procedure of collecting partial pressure of reactants and rate of formation of 2-MP and subjected these data to diffewrnt rate models based on surface reactions. From the derived rate equation, a tentative mechanism of the reaction has been suggested. An endeavour has also been made to support the present mechanism with the help of information to the author from literature.

II. Experimental

2.1 Catalyst Preparation

$MnFe_2O_4$ (MF) was prepared by low temperature, pH controlled co-precipitation route using dilute aqueous solution of metal salts. 1N NaOH solution was added to the solution containing 4.3g $Fe(NO_3)_3.9H_2O$ and 3.4g $MnCl_2.4H_2O$, under stirring till a pH of 8.5 was obtained. The precipitate was digested at 80°C for 2h. It was washed repeatedly with distilled water till free from chloride and nitrate ions. The catalyst was filtered, oven dried and calcined at 500°C for 6h at atmospheric pressure.

For preparation of alumina supported $MnFe_2O_4$, 16g of sized alumina (6/10 B.S.S. mesh size) was soaked in the dilute solution of manganese and iron salts (3.4 g of $MnCl_2.4H_2O$ and 4.3g of $Fe(NO_3)_3.9H_2O$ in 1 liter of water) for 3-4 hours. 1N NaOH was added to this system to allow precipitation of mixed hydroxides of the metals. The system was allowed to digest at 80°C for 4h. The catalyst was decanted and washed repeatedly with distilled water till free from chloride and nitrate ions. The material was oven dried and calcined at 500°C for 6h at atmospheric pressure to get alumina supported $MnFe_2O_4$ (AMF) catalyst.

2.2 Differential reactor

In this reactor, only a small amount of catalyst is used so as to keep the conversion level low. This permits direct evaluation of reaction rates. Because of small contact time, the composition remains practically constant throughout the catalyst bed and rates obtained are initial rates. The initial rates obtained under these condition are extremely helpful in simplifying rate equations. This technique is also helpful in dealing reactions with large heat effects.

The experimental set-up for kinetic measurement is similar to that described in chapter 2. 0.1 g of catalyst was used for kinetic studies.

2.3 Experimental procedure

The collection of data for kintic studies of the vapour phase cyclization of ED and PG were collected at atmospheric pressure in a vertical, down flow, fixed bed reactor. The upper half of the reactor worked as pre-heater and the lower half as the reactor. The fresh catalyst was charged in the centre of the reactor. Activation was attained by heating the catalyst in air at 773K for 4h and then cooling to the desired temperatures in a current of nitrogen and finally exposing to feed stream a mixture of ED, PG and water. Besides, functioning as a solvent, steam reduces the dealkylation of 2-MP to pyrazine, avoids charring, and reduces formation of aromatics. The liquid products were collected using a cold-water condenser. A Shimadzu 14B gas chromatograph equipped with flame ionisation detector and SE-30 column was used to determine the composition of product mixture. A blank run without any catalyst indicated negligible thermal conversion.

2.4 Identification and analysis of products

Before collecting data on a diferential reactor few experiments were performed in a macro-reactor taking 10 g of catalysts and feeding Ed, PG and water in the weight ratio9 of 1:1:2. About 50 ml of the product mixture was collected and was subjected to fractional distillation. The product boiling in the range 130-140°C was collected. A record of FT-IR spectrum of the distillate showed band in the region 2900-3200 cm^{-1}. Band appeared below 3000 cm^{-1} were assigned to C-H stretching modes of CH_3 group while those appeared above 3000 cm^{-1} were assigned to C-H stretching modes of ring hydrogen. There appeared medium to strong bands in the region 1000-1700 cm^{-1} characteristic of C-H / N-H bending modes and ring vibrations. These observations confirmed presence of 2-MP in the product. Further confirmation was made by comparing the FID retention time of 2-MP in the above mentioned distillate with that of standard 2-MP. Quantitative analysis was made on the basis of GLC peak area measurements. A Chemito model 7610 GLC machine was used for quantitative analysis. The analysis of product composition with differential reactor was made with GLC only.

2.5 Catalyst Characterization

The XRD records of alumina supported $MnFe_2O_4$ (AMF), $MnFe_2O_4$ (MF) were recorded over Rigaku X-ray powder diffractometer using Cu-K_α radiation as source. The recordings confirm the crystallinity of the samples and appearance of most of the peaks of support as well as catalyst (Culty et al., 1978). Ammonia desorption experiments were carried out to measure the acidity of the catalyst using ammonia as an adsorbate. Detailed procedure is described elsewhere (Radheshyam et al., 2002). The BET surface areas were measured by N_2 adsorption at liquid N_2 temperature using BET surface area analyzer (Model SAA-2002, S.P. Consultant, Mumbai) and was found to be M^2 g^{-1}. Scanning electron microscopy (SEM) pictures were obtained using JEOL JSM-5600 instrument. The subsequent elemental analysis was carried with the help of EDX. The FTIR spectrum of the alumina supported manganese ferrite catalyst was recorded on Perkin Elmer series 1600 FTIR spectrometer.

III. RESULTS AND DISCUSSION

3.1 Nomenclature

The reaction rates were calculated from the relation

$$R = F \cdot \frac{X_p}{S} \qquad 4.1$$

Where F is the flow rate of reactants, X_p, the percent composition of particular product in the reaction mixture as obtained from GLC and S is the surface area of the catalyst,

P_{EG} = partial pressure of ethylene glycol in
P_{ED} = partial pressure of ethylene diamine in
K_{EG} = adsorption equilibrium constant for adsorption of ethylene glycol
K_{ED} = adsorption equilibrium constant of ethylene diamine
k = rate constant of the reaction in.

3.2 Studies on effect of partial pressure of reactants

All kinetic measurements were performed under low conversion (bellow 10 % conversion pf PG). All measurements were taken after a study state condition was reached in the catalytic activity. Some standard experiments were performed from time to time to confirm that catalytic activity was constant and did not change with time. No reaction was observed at 350 °C in absence of catalyst even after several hours.

Effect of partial pressure of propylene glycol on rates

Effect of partial pressure of ethylene glycol on rates was studied by varying the same while keeping the partial pressure of ethylene diamine constant. Total pressure was kept constant by introducing some amount of inert gas in the feed. The data at 200 °C, 250 °C and 300 °C are presented in Table 4.1 and in Figures 4.1. It can be observed from the figures that the rate of rate of formation of 2-MP increases in sharply at lower partial pressure which tends to become constant at higher partial pressure.

Effect of partial pressure of ethylene diamine on rates

Effect of variation of partial pressure of ethylene diamines on rates was studied under constant partial pressure of propylene glycol and results are presented in Table 4.2 and figure 4.2. In this case also the rates increased at low partial pressure then tended towards a limiting value at higher partial pressure.

3.3 Effect of temperature on rates

Effect of temperature on rates is presented in Table 4.3 and in figure 4.3. As can be seen the Table 4.3 and Figute 4.3 rates increased with increase in temperature. The Arrehinius plot is shown in Fig. 4.4. The activation energy was calculated to be

3.4 Treatment of rate data

In the present study of the vapour phase kinetics of cyclization of propylene glycol and ethylene to 2-methylpyrazine over alumina supported mangenese ferrite catalysts, we collected data under conditions that conversion was bellow 10 %. Rate data collected under this condition can be termed as initial rates. Besides we used only 0.1 g of catalyst so as to minimise mass transfer and diffusional effects. Since mass transfer from gas phase to the catalyst surface and diffusion through the catalyst bed were not rate controlling, the possibility of surface adsorption of reactants and surface reactions as rate controlling steps were left. We tested two most popular models applied for such cases namely Langmuir-Hinselwood model and Reidel model. While farmer assumes reaction between adsorption pG as well as ED, the latter assumes reaction between a adsorbed reactant and another one remaing in gas phase. The rate laws derived on the basis of these two models are presented in Table 4.3. As the rate constants are physical constants their values can not be negative. An inspection of table 4.3 reveal that positive constants are found only for Reidel model assuming one reactant adsorbed on the surface and another one remaining in gas phase. A plot of observed rates along with calculated rates obtained from reidel model vs partial pressure are shown in Fig., and. Reasonable agreement between observed rates and calculated rates confirm the validity of Reidel model. Further confirmation of the model was obtained by ploting observed rates vs calculated rates. The plot is found to be a straight line with an inclination of 45 degree possing through origine. This again confirms the validity of the model.

3.5 Mechanism of the process

2-methyl pyrazine was found to be major product in the present study. Acetone was also detected in traces. When 2-methylpiperazine alone was passed over the catalyst in the reactor, 2-MP was found to be the only product. When PG alone was fed over the catalyst we obtained acetone as the major product. Passing hydrogen peroxide in the feed stream has no effect, indicating that free radical mechanism was not operative. Based on these observations, analysis of products and on the fact that kinetics follows a Reidel type of mechanism, an attempt is made here to presict a mechanism for the reaction.

Kulkarni have proposed a mechanism for synthesis of 2-MP from PG and ED over ZSM –5 catalyst. According to these authors, PG is adsorbed over the catalyst, gets protonated, looses a molecules of water and is converted into carbonium ion. The carbonium ion looses a proton and produces propylene oxide. Propylene oxide can react with ED to produce 2-methyl piperazine, which is dehydrogenated to produce 2-methyl pyrazine. Although, this sounds well, it does not support the Reidel type of mechanism in which PG is adsorbed on the surface and reacts with Ed present in Gas phase.

IV. Reaction Mechanism

The present catalyst contains both Lewis as well as Brönsted sites. While Lewis sites come from $MnFe_2O_4$, Brönsted sites come from hydroxyl group of alumina. It seems Lewis as well as Brönsted sites activate reaction of PG and ED. In case of Brönsted sites PG is adsorb first, gets protonated, looses a molecule of water and produces a carbonium ion. The carbonium ion looses a proton and produces propylene oxide. Propylene oxide can react with ED to produce 2-methyl piperazine, which is dehydrogenated to produce 2-methyl pyrazine. In order to confirm this we passed 2-methylpiperazine alone over the catalyst in the reactor, and obtained 2-MP as the only product. Propylene oxide can also rearrange to produce acetone. In fact, when PG alone was fed over the catalyst we obtained acetone as the major product. This mechanism is consistent with the mechanism proposed by (Kulkarni et.al, 1993) for synthesis of 2-MP from PG and ED over ZSM –5 catalyst. The mechanism over Brönsted sites is shown in Fig. 8(a). Over Lewis acidic sites above type of mechanism is rather unlikely, because of (1) absence of, Brönsted sites (2) high possibility of ED over PG for adsorption over Lewis Sites (ED is stronger base then PG). However, because of much higher concentration of PG & H_2O over ED, it is possible that they displace few adsorbed ED molecules and produce Brönsted sites. Once such Brönsted sites are produce, the mechanism will follow the usual route of PG protonation, dehydration, formation of propylene oxide and reaction with ED to produce 2-methyl piperazine followed by its dehydrogenation to give 2-methyl pyrazine. This mechanism is consistent with the mechanism proposed by Forni and Paolo (Forni & Paolo, 1991) in their studies on TPD–TPR–MS Mechanistic study of the synthesis of 2-methylpyrazine over palladized Zn–Cr Oxide. The mechanism over Lewis sites is shown in Fig. 8(b) When ED alone was fed over the catalyst ammonia and pyrazine were obtained. The mechanism of by products formation is shown in Fig. 8(c).

References

1. Interrante, L. V., Hampden-Smith, M. J., (Editors) "Chemistry of Advanced materials: an overview", Wiley-VCH, New York (1998).
2. Anpo, M., Takeuchi, M., "The design and development of highly reactive titanium oxide photocatalysts operating under visible light irradiation". J. Catal., Vol. 216, No. 505-516 (2003).
3. Maira, A. J, Yeung, K. L., Lee, C. Y., Yue, P. L., Chan, C.K., "Size effects in gas-phase photo-oxidation of trichlorethylene using nanometer-sized TiO_2 catalysts", J. Catal. Vol. 192, No. 185-196 (2000)
4. Sterba, M.J. Haensel, V., "Catalytic reforming". Ind. Eng. Chem. Prod. Res. Dev., Vol. 15 NO. 2-17 (1976)
5. Forni, L., Pollesel, P., "TPD-TPR-MS mechanistic study of the synthesis of 2-methylpyrazine over palladized Zn-Cr oxide" J. Catal. Vol. 130 No. 403-410 (1991).
6. Kulkarni, S.J., Subramanyam, M., Rama Rao, A.V., "Intermolecular and intramolecular cyclization over ZSM-5 and chromite catalysts to synthesize 2-methyl pyrazine and piperazine", Indian J. Chem, Vol. 32A, No. 28-32 (1993).
7. Ilnam, P, Jeongo, L., Youngoo, R., Yohan, H., Hyungrok, K. "$CuO/ZnO/SiO_2$ catalysts for cyclisation of propylene glycol with ethylene diamine to 2-methyl pyrazine". Appl. Catal. A, Vol. 253 No. 249-255 (2003).
8. Balpanov, D. S., Krichevskii, L. A., Kagarlitskii, A. D., "Binary oxide systems in catalytic synthesis of 2-methyl pyrazine from 1, 2-propylene glycol and ethylenediamine". Russian J. Appl. Chem. Vol. 74, No. 2065-2067 (2001).
9. Anand, R., Hegde, S. G., Rao, B. S., Gopinath, C. S., "Catalytic synthesis of 2-methyl pyrazine over Zn-modified zeolites". Catal. Lett. Vol. 84, No. 265-272 (2002).
10. Anand R., Rao, B. S., "Synthesis of 2-methyl pyrazine over zinc-modified ferrierite (FER) catalysts". Catal. Comm. Vol. 3, No. 29-35 (2002).
11. Subramanyam, M., Kulkarni, S. J., Rama Rao, A. V., "Catalyst preparation studies for the synthesis of 2-methyl pyrazine". Ind. J. Chem. Vol 2, No. 237-240 (1995).
12. Cullity, B. D., "Elements of x-ray diffraction", 2nd edn. Addison-Wesley Publishing Co. New York, pp 326 (1978).

13. Radhe Shyam, A., Dwivedi, R., Reddy, V. S., Chary, K.C.R., Prasad, R., "Vapour phase methylation of pyridine with methanol over the $Zn_{1-x}Mn_xFe_2O_4$ (x = 0, 0.25, 0.50, 0.75 and 1) ferrite system". Green Chemistry Vol 4, No. 558-561 (2002).
14. Ladgaonkar, B. P., Kolekar, C. B., Vaingankar, A. S., "Infrared absorption spectroscopic study of Nd^{3+} substituted Zn-Mg ferrites". Bull. Mater. Sci., Vol. 25, No. 4, 351-354 (2002).

Table 4.1 : Effect of patial pressure of propylene glycol on rates Weight of catalyst : 0.1g Partial pressure of propylene glycol : 0.04-0.20 Partial pressure of ethylene diamine :0.22 Rate x 10^6 (moles min^{-1} m^{-2})

S.No.	Partial pressures of PG	Temperature					
		200 °C		250 °C		300 °C	
		Observed	Calculated	Observed	Calculated	Observed	Calculated
1	4	2.11	2.11	17.70	17.70	35.97	35.96
2	8	3.60	3.60	29.38	29.38	60.88	60.88
3	12	4.72	4.72	37.67	37.67	79.15	79.15
4	16	5.58	5.58	43.86	43.86	93.12	93.12
5	20	6.23	6.23	48.65	48.65	104.5	104.15

Table 4.2 : Effect of patial pressure of ethylene diamine on rates Weight of catalyst : 0.1g Partial pressure of propylene glycol : 16x10-2 Partial pressure of ethylene diamine :9.0-27.2 x 10^{-2} Rate x 10^6 (moles s^{-1} m^{-2})

S.No.	Partial pressures of ED	Temperature					
		200 °C		250 °C		300 °C	
		Observed	Calculated	Observed	Calculated	Observed	Calculated
1	9.0	3.1	3.1	22.65	22.65	46.11	46.11
2	13.6	4.21	4.21	31.32	31.22	64.85	64.85
3	18.1	5.01	5.01	38.49	38.49	80.83	80.83
4	22.7	5.67	5.67	44.75	44.75	95.19	95.19
5	27.2	6.20	6.00	50.06	50.06	108.14	108.14

Table 4.3 : Reaction rate constant and adsorption equilibrium constantss

Temperature(K)	I/T	K	K_A	K_B
473	-4.130	0.000074	38.61	29.51
523	-3.107	0.00077	22.61	11.41
573	-1.721	0.0019	12.49	5.68

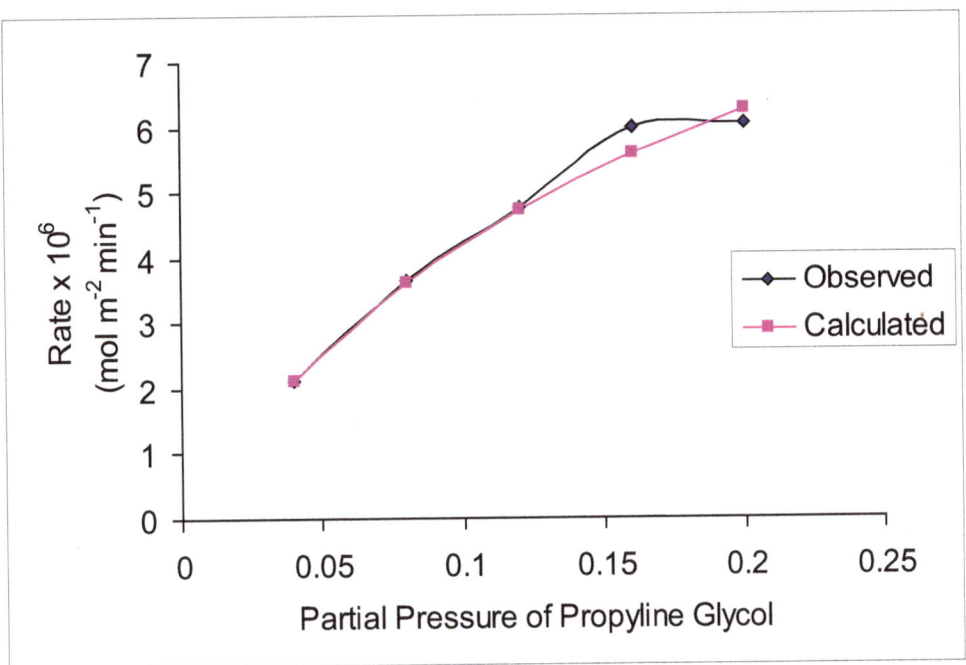

Fig.1 : Effect of Pp of PG on rate of formation of 2 - MP Temp. 200 °C

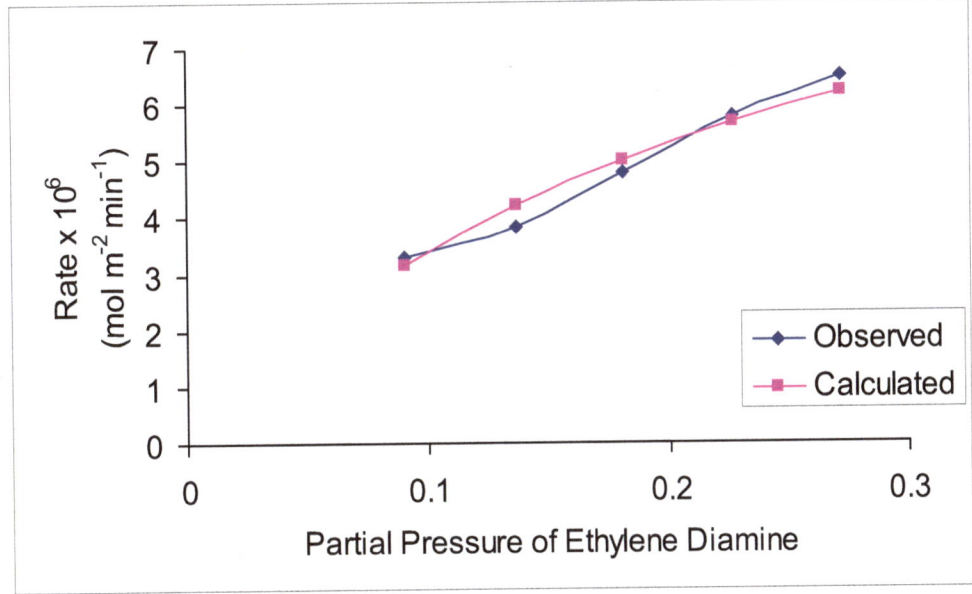

Fig. 2 : Effect of Pp of ED on rate of formation of 2 - MP Temp. 200 °C

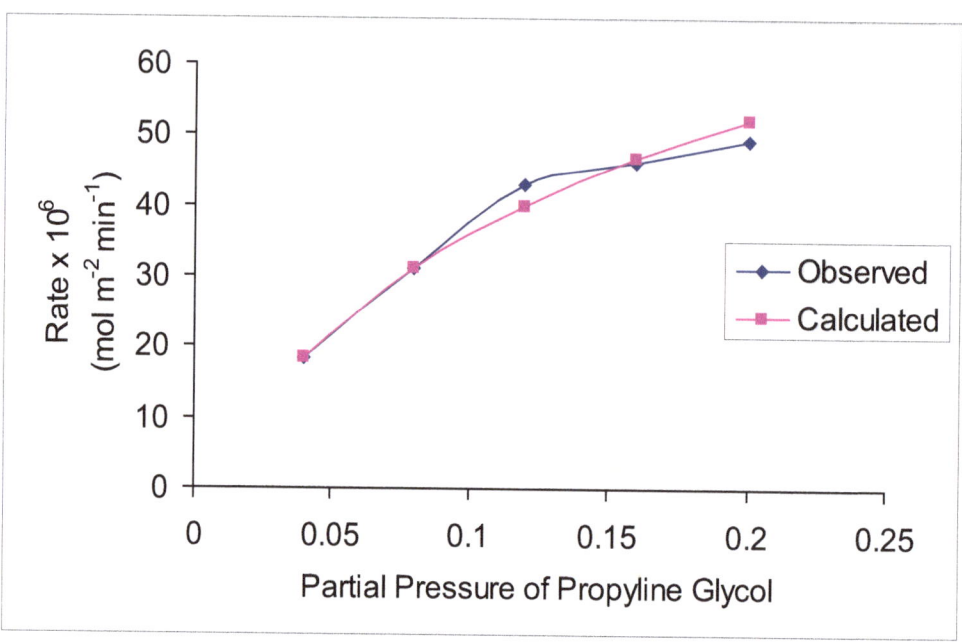

Fig. 3 : Effect of Pp of PG on rate of formation of 2 - MP Temp. 250 °C

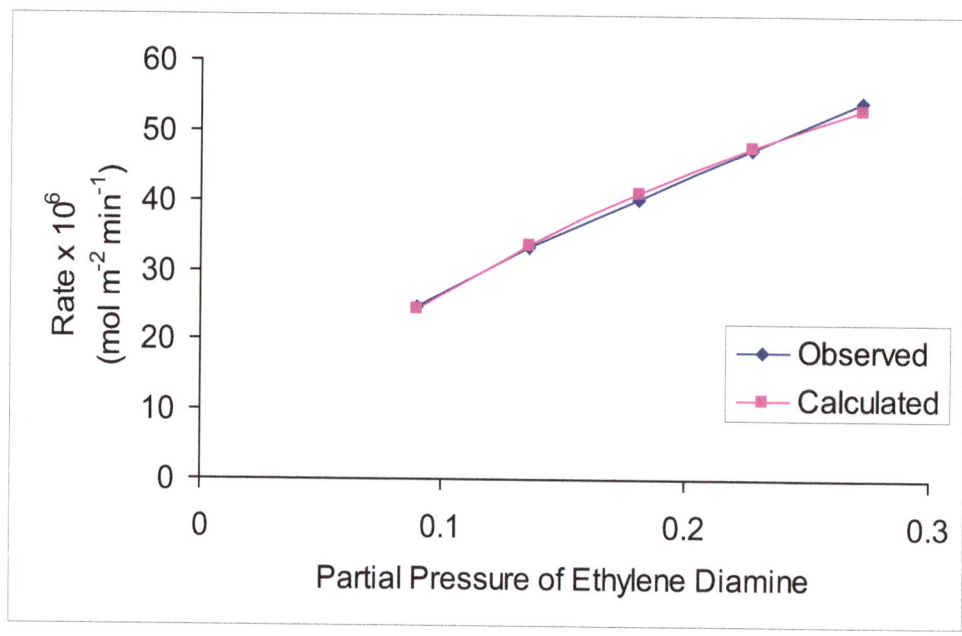

Fig. 4 : Effect of Pp of ED on rate of formation of 2 - MP Temp. 250 °C

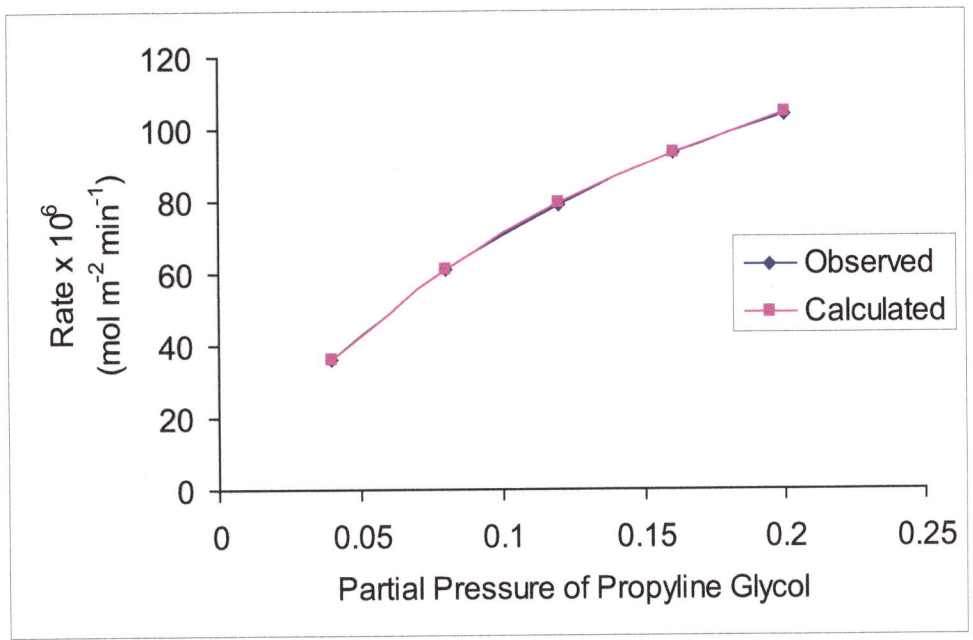

Fig.5 : Effect of Pp of PG on rate of formation of 2 - MP Temp. 300 °C

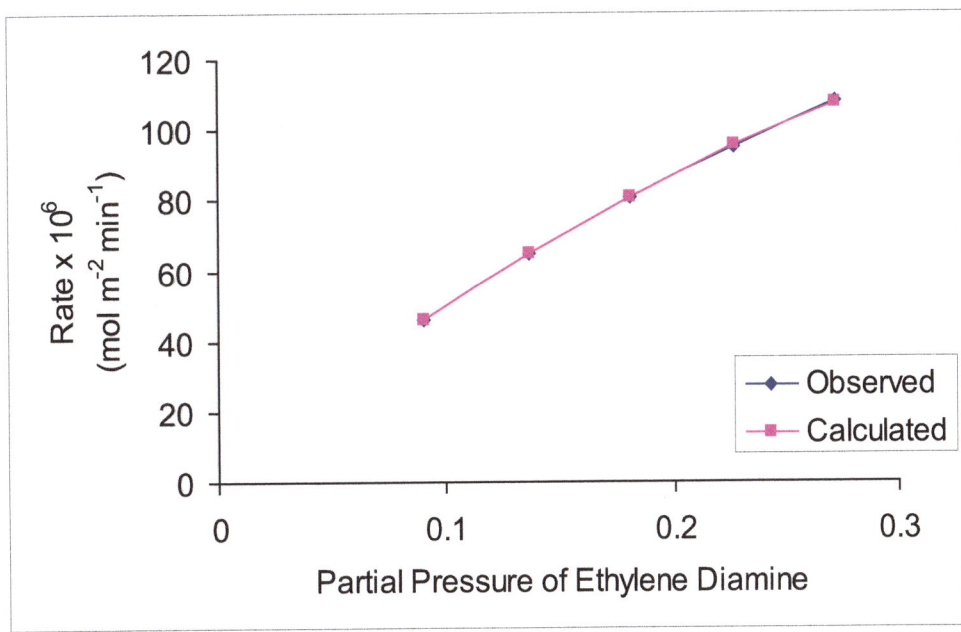

Fig. 6 : Effect of Pp of ED on rate of formation of 2 - MP Temp. 300 °C

Figure : Reaction mechanism of vapour phase catalytic synthesis of 2-MP over alumina Supported manganese ferrite catalysts (a) bronsted sites (b) Lewis sites (c) formation of by products

Chapter - 5

Kinetics Studies and Mechanism Evolution of the Methylation of Phenol over AlFe$_2$O$_4$ Catalyst

I. Introduction

Alkylation of phenol with methenol is an industrial important reaction since alkyl phenols such as o-cresol and 2,6 xylenol are used as raw material for the synthesis of commercially important products [1]. Zeolites [2-13], metal phosphates [14,15], hydrotalcites [16] and ferrites [17, 18] have been used as catalysts. While zeolites lead to O-alkylated as well as C-alkylated products we have noted that ferrites produce only ortho-C-alkylated products such as o-cresol and 2,6-xylenol.

There has been an upsurge on the alkylation of organic compounds over ferrospinel catalysts [19-21], having general formula AB$_2$O$_4$. Many ferrites have the spinel type structure which can be described in terms of a nearly cubic close packed arrangement of anions with one half of the octahedral interstices (B site) and one eighth of the tetrahedral interstices (A site) filled with cation. Most of the ferrite properties depend upon the distribution of metal ions in the tetrahedral and octahedral sites within the spinel structure. The adequate selection of the substitution ion and the appropriate chemical composition changes the catalytic properties of the ferrospinels [22]. A thorough literature survey on phenol methylation reveals that most of the published work on methylation of phenol. is on process development. A kinetic study of the vapour phase methylation of phenol over magnesium~aluminium <ialcined hydrotalcites is reported by Velu and Swamy [16]. A differential equation base on plug flow reactor was fitted to the observed data.

Depending upon the position of metals in the tetrahedral and octahedral sites ferrospinel can be normal $A^{+2}[B_2^{+3}]O_4$, inverse $A^{+3}[A^{+2}B^{+3}]O_4$ or mixed spinel in which the divalent cations are distributed between both sites. This type of cation distribution significantly affects acido-basic and surface properties of ferrospinels [23,24]. Although there are a couple of reports on the kinetics of aniline alkylation over industrial niobic acid [25], Zn-Co ferrite [26] and zeolite [27] catalysts, to the best of our knowledge, there is no report on the kinetics and mechanism of this reaction over Al Ferrite in vapor phase. The present study was therefore undertaken with a view to (1) collect data on the kinetics of the vapour phase alkylation of phenol with methanol over AlFe$_2$O$_4$ catalyst (2) to find a suitable rate law, which can explain the data satisfactorily, and (3) to predict a mechanism of the reaction.

II. Experimental

a) Catalyst preparation

AlFe$_2$O$_4$, were prepared by low temperature, pH controlled co-precipitation route using aqueous solutions of ferric nitrate & Aluminium chloride. NaOH was added to the mixture solution under stirring till a pH of 8.5 was obtained. The precipitate was washed, filtered, dried at 383 K and calcined at 873 K for 6 hrs in a current of Air.

a) Characterization

b) Apparatus and procedure

Kinetic experiments were performed using a fixed bed down flow Pyrex glass tubular differential reactor (0.35 m in length and 0.8cm ID) at atmospheric pressure. In this reactor, only a very small amount of catalyst is employed so as to keep the conversion level low, and this affords direct evaluation of reaction rates. The contact time being small, the gas composition remains practically constant throughout the catalyst bed. The rates thus obtained are the initial rates. These initial rates are extremely helpful in simplifying the rates equations to make their applications In addition; this technique appears to be the last resort in cases of reactions with large heat effect. The reactant partial pressure varied in the ranges as follows: Phenol (P_a) 4.15 – 29.08 kPa and methanol (P_m) 1.41 – 9.92 kPa. The measurements were made at three temperatures namely 543, 593 and 643 K. A 10 cm^3 pressure-equalizing funnel fed the mixture of phenol and Methanol. The liquid products were condensed with the help of a cold-water condenser a cold trap and were analyzed by Shimadzu 14B Gas Chromatograph using SE-30 column and FID detector.

c) Removal of Diffusional and Mass transfer Effects

A plot of total rate of formation of xylenol against particle size is shown in Fig. 4. The rate of reaction gradually increases with decreasing particle size and ultimately become constant when particles are in the range

0.5-0.8 mm. It can be concluded that with particle in the range 0.5-0.8 mm. diffusional effects are negligible. Therefore further studies were conducted with catalysts particles of 0.5 mm.

A series of Experiment were conducted at different gas flow rates keeping the space velocity constant by suitable adjustment of catalyst volume. A plot of catalyst bed volume against rate reveal that bulk diffusion will be negligible at weight hour space velocity of $0.05h^{-1}$.

d) Kinetic models

Since diffusion through the catalyst surface, mass transfer from the gas stream to the catalyst surface and desorption of products were not rate controlling, the possibility of adsorption of reactants and surface reaction as rate controlling was left. Several rate equations based on adsorption and surface reaction were derived. Such equations were solved for rate constants and adsorption equilibrium constants. A summary of different mechanisms and rate expressions derived on the basis of these is presented in the Table 1. The mechanism for models 1 is the reaction between adsorbed phenol and adsorbed methanol on the catalyst surface and the slow step is the surface reaction. The mechanism of model 2 assumes the reaction between adsorbed methanol and phenol in gas phase as the slow and the rate-controlling step. These two models are based on Langmuir adsorption as described elsewhere [28,29]. Model 1 can be referred to as the Langmuir- Hinshelwood model, while model 2 is the familiar Rideal-Eley model.

III. Result and Discussion

a) Effect of partial pressure of methanol

The effect of partial pressure of methanol on the rates was studied at constant partial pressure of phenol and results are represented graphically in Fig. 7. The rates initially increased at low partial pressure of methanol and gave indications of a limiting rate being attained at higher partial pressure.

b) Effect of partial pressure of Phenol

The effect of variation of partial pressure of phenol on the rate of formation of xylenol was studied at constant partial pressure of methanol. The results are shown graphically in Fig. 8. The rates initially increased linearly with partial pressure of phenol and attained a limiting value at higher partial pressure.

c) Effect of Temperature

The Effect of temperature on the rates was investigated between 543-643 K and at various partial pressures of phenol and Methanol. The results are presented in Table 2 and Figs. 7 & 8. As is evident, the rates increased with increase temperature.

IV. Criteria for Discriminating between Different Models

Since the constants involved in models 1 & 2 are physically constants, there value should be positive. Hence, based on rate equations shown in Table 3, the rate or adsorption coefficients were evaluated by method of least squares and the mechanisms giving a negative coefficient in the rate equation were rejected. Positive coefficients were found only for model 2. The correlation of data is illustrated in Fig. 7 & 8 where observed rates are plotted along with calculated rates for different partial pressures of phenol and methanol and in Fig. 9 where observed rates are plotted against calculated rates. A plot of residuals versus calculated rates is shown in Fig. 10. The plot does not show any heteroscedic pattern.

V. Mechanism

Adsorption of phenol on the surface can be either horizontal [30] or vertical [21]. Horizontal absorption is expected to promoted oxygen and para substituted products while vertical adsorption will promote ortho substitute products. In the present study major products were ortho substituted with negligible are no oxygen and para substituted products. This suggest only vertical adsorption by phenol and rules out horizontal adsorption. It seems phenol is vertically adsorbed on the surface with oxygen attached to a Lewis acid site and hydrogen with the neighboring Lewis base site. Methenol adsorbed on the neighbroing Lewis base site abstract a proton from phenol, get protoneted and converted to carbonium ion by releasing water molecule. An electrophilic attack by carbonium ion on the ortho position of phenoxide ion leads to formation of o-cresol and completing the catalytic cycle. Further adsorption by o-cresol and methenol leads to the formation 2,6-xylenol. The alkylation of phenol is a consecutive reaction on the catalyst surface. The presence of negative charge on the ion requires a strong acidic site for its adsorption. This seems to be the reason for increase of orth products.

In phenol alkylation major product is Ortho di-substituted on the other hand in aniline alkylation is mono Nitrogen substituted compound. Answer to this perhaps lies in the stereochemistry of transition state. Phenoxide ions provides equal opportunity to both Ortho position for electrophilic attack on the other hand bulky phenil and ethyl group and mono ethyl aniline, hinders its adsorption on the surface and suppresses formation of di-substituted product.

VI. Conclusions

Kinetics of alkylation of phenol with methanol has been studied over Al ferrite catalyst in the temperature range 543-643 K. A rate law based on Rideal-Eley model satisfied the observed data. The activation energy has been computed to be 25.3 Kcal. mol^{-1}. A mechanism for the process has been proposed.

Références

1. Ullmann's Encyclopedia of Industrial Chemistry, vol. 9, sixth ed., Wiley- VCH, Weinheim, 2003, p. 642.
2. P.D. Chantal, S. Kaliaguine, J.L. Grandmaison, Appl. Catal. 18 (1985) 133.
3. R. Pierantozzi, A.F. Nordquist, Appl. Catal. 21 {1986) 263.
4. S.C. Lee, S.W. Lee, K.S. Kim, T.J. Lee, D.H. Kim, J. Chang Kim, Catal. Today 44 (1998) 253.
5. M. Marczewski, J.-P. Bodibo, G. Perot, M. Guisnet, J. Mol. Catal. 50 (1989)211.
6. L. Garcia, G., Giannetto, M.R. Goldwasser, M. Guisnet, P. Magnoux, Catal. Lett. 37 (1996) 121.
7. S. Balsama, P. Beltrame, P.L. Beltrame, P. Carniti, L. Forni, G. Zuretti, Appl. Catal. 13 (1984) 161.
8. P. Beltrame, P.L. Beltrame, P. Carniti, A. Castelli, L. Fomi, Appl. Catal. 29 (1987) 327.
9. R.F. Parton, J.M. Jacobs, H. van Ooteghem, P.A. Jacobs, Stud. Surf. Sci. Catal. 46 (1989) 211.
10. S. Namba, T. Yashima, Y. Itaba, N. Hara, Stud. Surf. Sci. Catal. 5 (1980) 105.
11. Z.-H.Fu, Y.Ono, Catal.Lett. 21 (1993) 43.
12. G. Moon, W. Bohringer, C.T. O'Connor, Catal. Today 97 (2004) 291.
13. M.D. Romero, a. Ovejero, A. Rodriguez, J.M. Gomez, I. Agueda, Ind. Eng. Chem. Res. 43 (2004) 8194.
14. G. Sarala Devi, D. Giridhar, B.M. Reddy, J. Mol. Catal. 181 (2002) 173. [15]F.M. Bautista, J.M. Campelo, A. Garcia, D. Luna, J.M. Marinas, A. Ro-mero, J.A. Navio, M. Macias. Appl. Catal. A 99 (1993) 161.
15. R. Bal, B.B. Tope, S. Sivasanker, J. Mol. Catal. A 181 (2002) 161. [17] S. Velu,C.S. Swamy, Appl. Catal. A 145 (1996) 141.
16. V.V. Rao, V. Durgakumari, S. Narayanan, Appl. Catal. 49 (1989) 165. [19] S. Velu, C.S. Swamy, Appl. Catal. A 162 (1997) 81.
17. K. Sreekumar, T. Raja, B. P. Kiran, S. Sugunan , B. S. Rao, App Catal A. (1999) 182-327
18. K. Sreekumar, T. M. Jyothi, T. Mathew, M. B. Talawar, S. Sugunan , B.S. Rao, J.Mol. Catal (2000) 159: 327-334.
19. K. Sreekumar,T Mathew, Devassy B M, Rajgopal R, Vetrivel R and Rao B S, Appl. Catal A. 205 (2001) 11-18
20. V. S. Reddy, A.Radheshyam, R.Dwivedi, R.K. Gupta, V. R. Chumbhale, R. Prasad, J. chemi. technol.& biotechnol. 79 (2004) 1057-1064.
21. J. P. Jacobs, A. Maltha, J. G. H. Reintjes, J. Drimal, V. Ponec and H. H. Brongersma, J. catal. 147 (1994) 294-300.
22. C. G. Ramankutty , S. Sugunan, Appl. Catal A 218 (2001)1039-1051
23. B. Frank, D. Habel, R. Schomacker, Catalysis letters 100 (2005) 181- 187.
24. K. Sreekumar, T.Jyothi, R.C.Govindankutty, B.S Rao, S. Sugunan, Reaction Kinetics and Catalysis Letters. 70 (2000) 161- 167.
25. W.O. Hagg, P.B. venuto., Adv Chem. Ser. 102 (1971) 260.
26. B.D. Cullity, element of X-ray Diffrection, 2nd Edn, Addison-Wesley publishing Co. (1998) 326.
27. A. Radheshyam, R. Dwivedi, V. S. Reddy, K.V.R. Chary, R. Prasad, Green Chemistry 4 (2002) 558-561.
28. J.M. Smith, Chemical Engineering Kinetics, 2nd edn. McGraw- hill, New York, 1970.

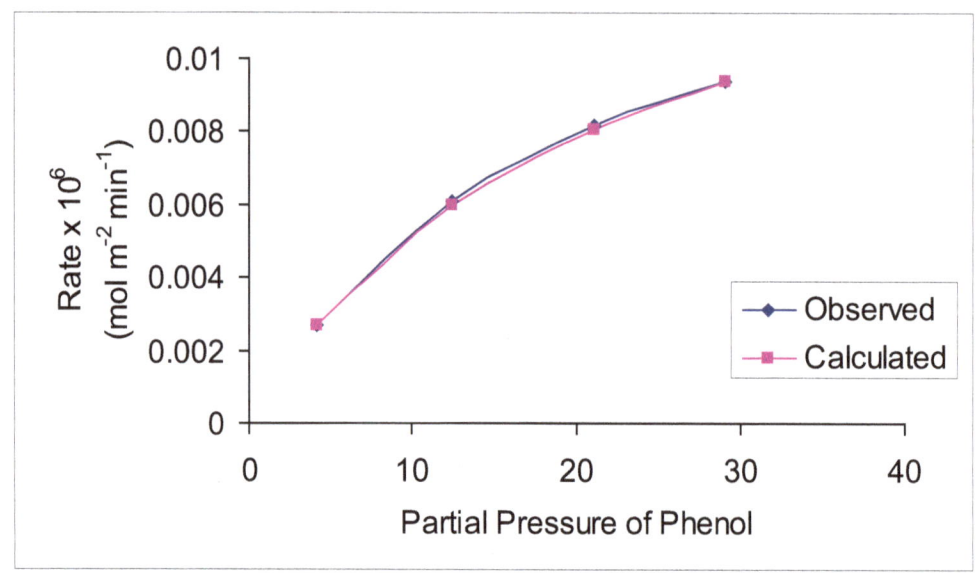

Fig. 1 : Effect of Pp of Phenol on rate of formation of Xylenol Temp. 270 °C

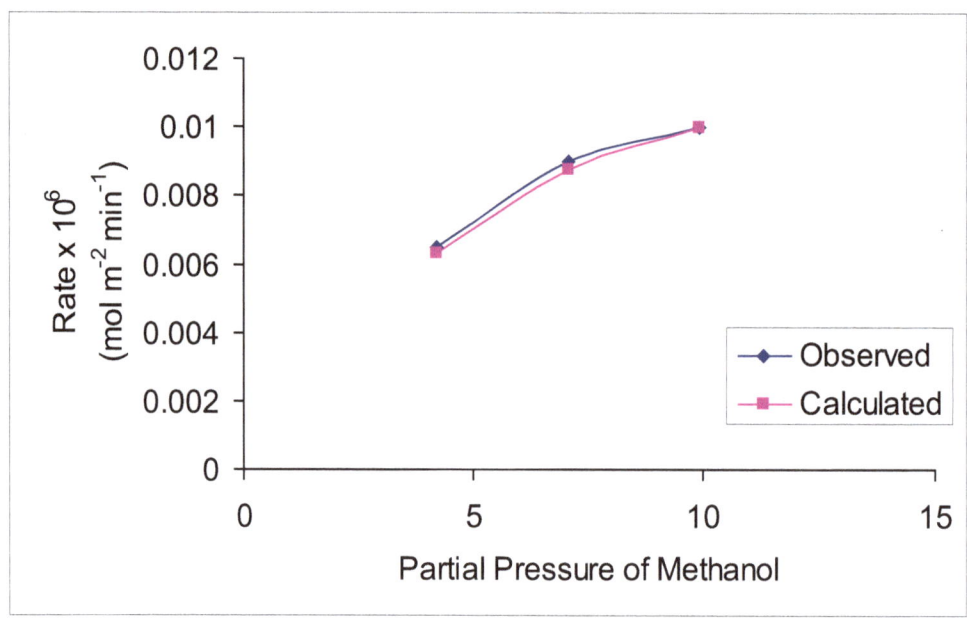

Fig. 2 : Effect of Pp of Methanol on rate of formation of Xylenol Temp. 270 °C

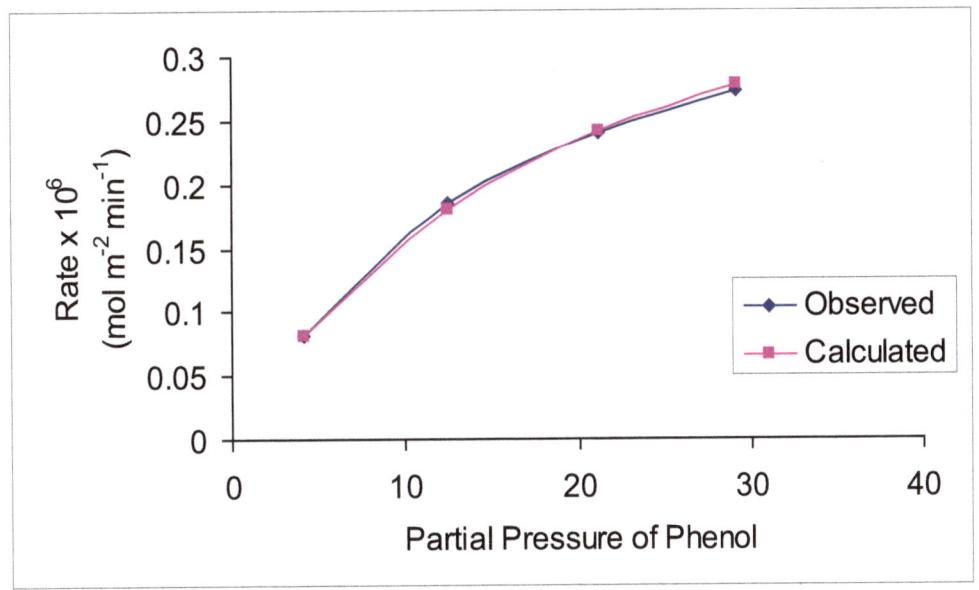

Fig. 3 : Effect of Pp of Phenol on rate of formation of Xylenol Temp. 320 °C

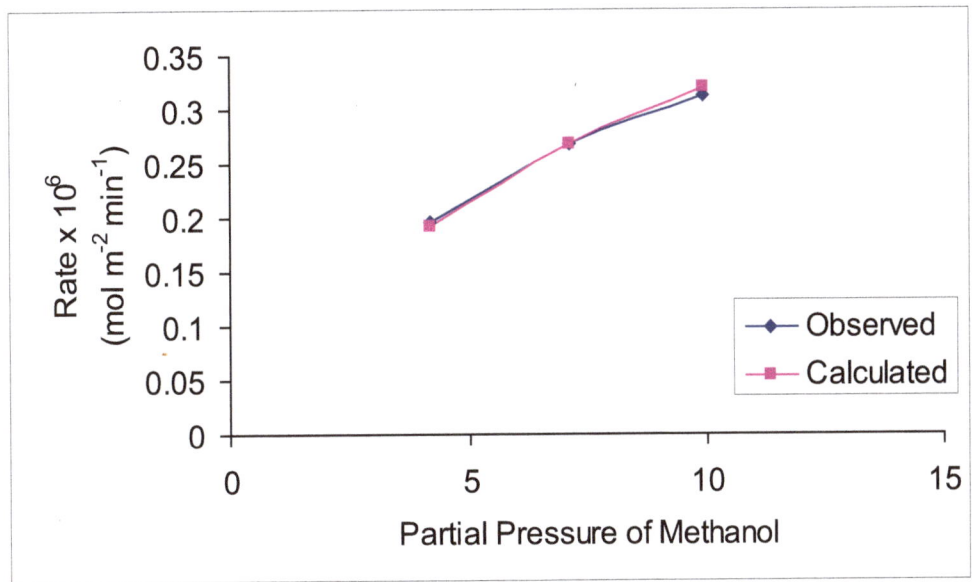

Fig. 4 : Effect of Pp of Methanol on rate of formation of Xylenol Temp. 320 °C

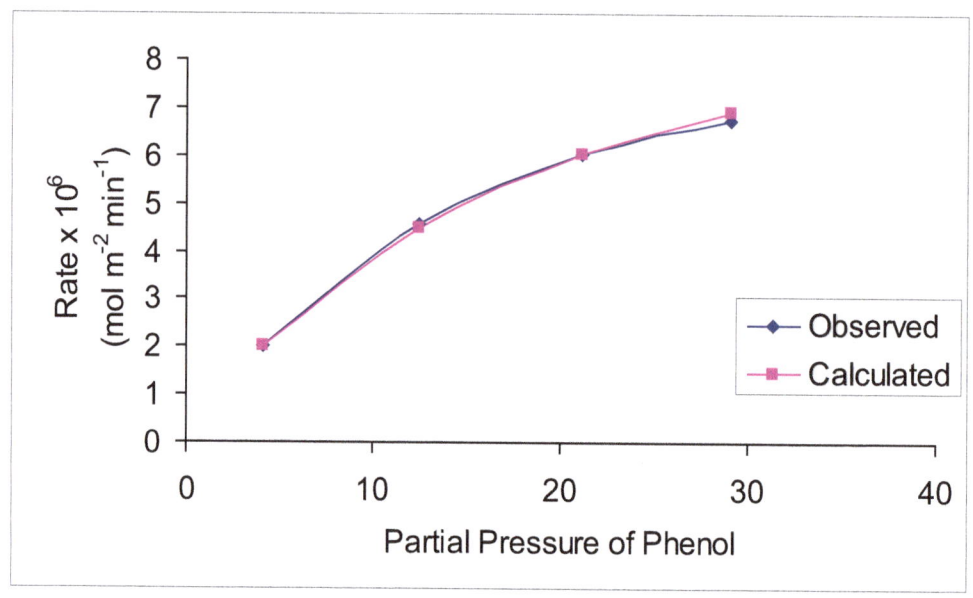

Fig. 5 : Effect of Pp of Phenol on rate of formation of Xylenol Temp. 320 °C

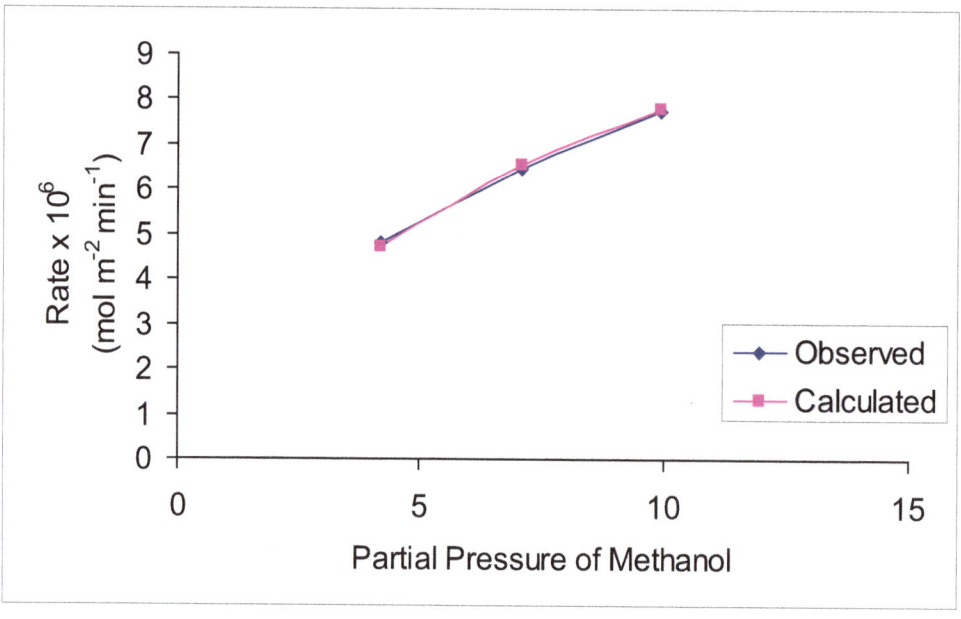

Fig. 6 : Effect of Pp of Methanol on rate of formation of Xylenol Temp. 370 °C

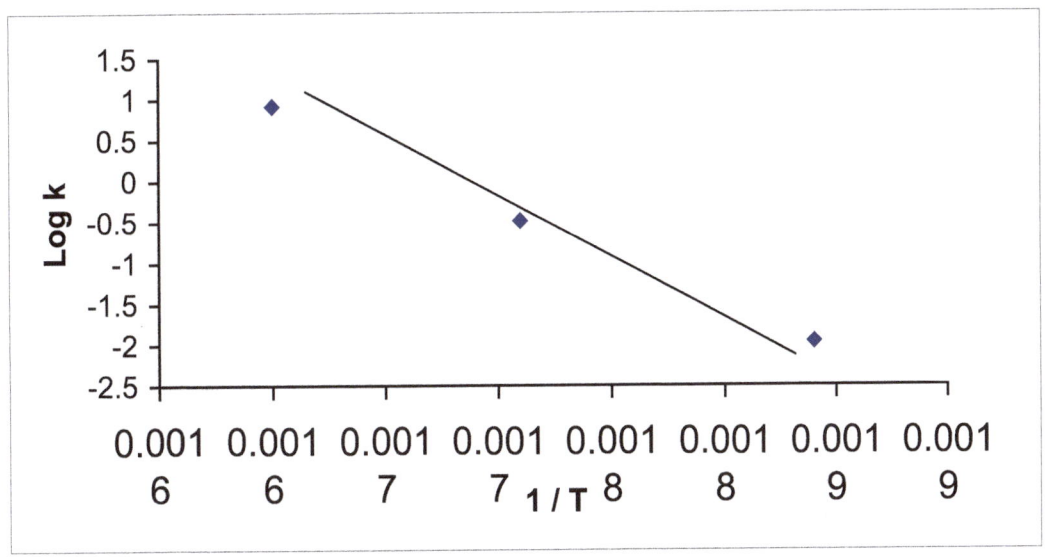

Fig. 7 : Rate of formation of xylenol vs. Temperature

Table 1 : Models Tested for alkylation of Phenol with methanol

Model	Mechanism	Resulting rate equation
1.	**Langmuir-Hinshelwood model** ROH + S → ROH–S CH_3OH + S → CH_3OH–S	$R = k K_A K_M P_A P_M / (1 + K_A P_A + K_M P_M)^2$
2.	ROH–S + CH_3OH–S → R–CH_3(OH) + H_2O **Rideal-Eley Model** CH_3–OH + S → CH_3–OH–S CH_3–OH–S + ROH → CH_3–R(OH) + S CH_3–R(OH) + CH_3OH–S → R–OH(CH_3)(CH_3) + S	$R = k K_M P_A P_M / (1 + K_M (P_M))$

* R = C_6H_5 , S = Surface, P_M = Partial pressure of Methanol, P_A = Partial pressure of Aniline, K = Equilibrium Constant, k = Rate Constant

Table 2 : Effect of Partial pressure of Phenol on the rates

S.No.	Partial pressures of Phenol	Temperature					
		270 °C		320 °C		370 °C	
		Observed	Calculated	Observed	Calculated	Observed	Calculated
1	4.15	.0027	0.0027	0.0812	0.0815	2.0112	2.0168
2	12.46	.0061	0.006	0.1861	0.181	4.5801	4.497
3	21.15	.0082	0.0081	0.2402	0.242	6.0302	6.01471
4	29.08	.0094	0.0094	0.2734	0.278	6.7633	6.929

Table 3 : Effect of Partial pressure of Methanol on the rates

S.No.	Partial pressures of Methanol	Temperature					
		270 °C		320 °C		370 °C	
		Observed	Calculated	Observed	Calculated	Observed	Calculated
1	4.21	0.0065	0.0063	0.1965	0.1931	4.7966	4.7374
2	7.09	0.009	0.0088	0.269	0.268	6.4764	6.5427
3	9.92	0.01	0.01	0.3122	0.32	7.7455	7.7789

Table 4 : Reaction rate Constants and activation Energy

Constants	Temperature (K)			Activation Energy (k cal / mol)
	543	583	623	
k	0.011455	0.331534	8.265109	25.3
K_A	0.137596	0.133970	0.140272	
K_B	0.057930	0.060517	0.060405	

Reaction Mechanisms:

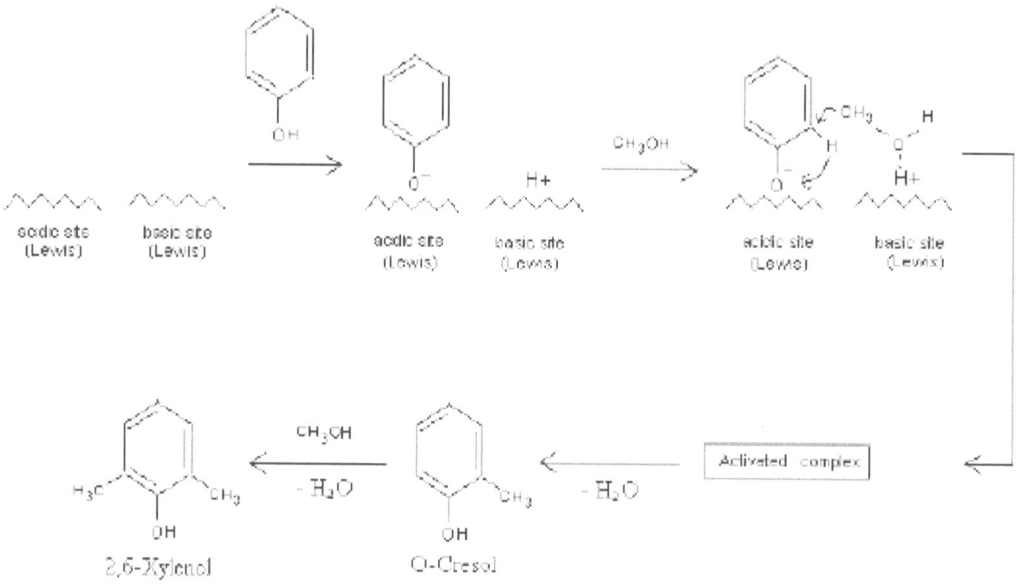

Scheme-1 Vertical orientation

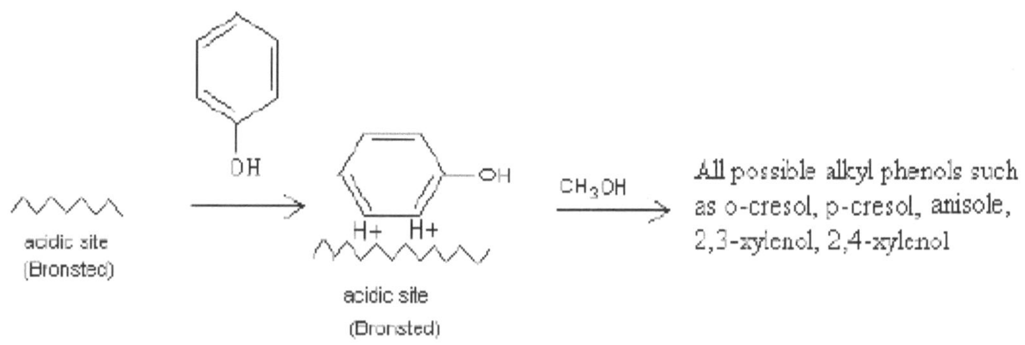

scheme-2 Horizontal orientation

Scheme 1 : Mechanism of the alkylation of phenol with methanol over $AlFe_2O_4$ catalyst

INDEX

A

Acetone · 16
Alkyl · 5, 33, 48
Arrehinius · 16

B

Bronsted · 33, 34, 53, 54

C

Catalytic · 1, 6, 17, 37, 38, 47
Chemitogram · 1
Coercivity · 5, 13

D

Diatomaceous · 44, 45
Distillation · 15, 35, 45

F

Ferrite · 1, 13, 24, 50
Ferrospinels · 7, 24, 35, 47

H

Heteroscedic · 25

I

Ionisation · 15
Isotherm · 44, 58

K

Kinetics · 13, 24, 26, 33
Kirichenko · 48, 51

L

Langmuir · 16, 25, 30, 34, 42, 48
Lorentzian · 6

M

Methylation · 5, 24
Monochromator · 2
Mössbauer · 2, 13, 36, 41, 42

N

N-Butylaniline · 47

P

Preheater · 1
Propylene · 1, 13, 16, 17, 18, 33, 34
Pyrazinamide · 13, 34, 49

Q

Quadrupole · 6, 42

S

Shimadzu · 6

T

Tertbutyl · 48
Tetrahedral · 5, 6, 7, 24, 36, 38

Z

Zeolites · 5, 13, 17, 24, 33, 34, 47, 49

Knowing without seeing is at the heart of chemistry.

— *Roald Hoffmann* —

ISBN 978-1-5330-2549-4

www.ingramcontent.com/pod-product-compliance
Lightning Source LLC
Chambersburg PA
CBHW050858180526
45159CB00007B/2714